GCSE PASSBOOK

PHYSICS

Barry Stone

First published 1988
by Charles Letts & Co Ltd
Diary House, Borough Road, London SE1 1DW

Illustrations: Illustrated Arts Ltd.

British Library Cataloguing in Publication Data
Stone, Barry
    Physics. – (Key facts, GCSE passbooks).
    1. Physics
    I. Title   II. Series
    530   QC23

    ISBN 0 85097 802 5

Printed and bound in Great Britain by
Charles Letts (Scotland) Ltd

Contents

GCSE is an exciting development in education – a welcome revolution. It will make the teaching/learning of experiences especially rewarding in the sciences, particularly in Physics.

The task of preparing this text has been a long but pleasant one. I am indebted to Josephine Fageant for her keyboard skills and to Trevor Hook for his invaluable advice in the preparation of the manuscript. I am also most grateful to Diane Biston and her colleagues at Charles Letts & Co Ltd for their help. Perhaps most of all, my thanks go to the teaching staff of Porchester School, Bournemouth, who (so long ago!) encouraged me in an enthusiasm for science.

This book is unlike many Physics books. It does not aim to 'teach' or try and teach Physics. Its purpose is to aid revision. The new GCSE courses have been specially written and this book is designed to follow them.

Any Physics syllabus designed by the examination boards must include the Core Content for GCSE. This book covers the Core Content. In addition, the individual boards use other material/ topics in their syllabus. This book covers the major additions for all the boards. Any student studying Physics for GCSE should therefore find this book suitable for their particular syllabus.

This is not a 'questions' book, but an aid to revision. There are questions at the end of each chapter, but they are only for revision and guidance. It will also serve as a useful and handy reference book. Each chapter has been arranged into four parts;

1 Aims These give you an idea of what you should be able to do at the end of the chapter. Treat them as suggestions.

2 Contents Each major area of the syllabus has been broken down into sections. The material follows a logical order. Key words are emphasized and a special 'K' symbol is used to indicate particularly important ideas.

3 Questions These are to give you practice in answering questions in an 'examination style' and not for testing your in-depth knowledge of the subject.

4 Summary This section refreshes your memory on the main ideas contained in the chapter.

The layout of the book follows a logical pattern and each of the chapters could be worked through in order, each new chapter building on the previous one. For example, Chapter 6 Electronics would best be revised after Chapter 5 Electricity.

Any GCSE syllabus in Physics must include the minimum core of content set out below.

1 Matter: structure of an atom; electric charge; radioactivity; particulate nature of matter; kinetic theory model of matter; states of matter; basic electrical, thermal and structural properties of matter.
2 Energy: energy forms; principles of the conservation of energy; changing energy from one form to another; work and power; energy transfer, waves and wave motion; electromagnetic spectrum.
3 Interactions (happenings): forces and their effects on size and motion; balanced forces; electrical; electromagnetic.
4 Physical quantities: units (mass, length, time, temperature and current); speed, acceleration, force, density, pressure, potential difference and resistance.
5 Applications of Physics: the application of Physics should be generously illustrated/demonstrated throughout any Physics course.

Each of the examination boards has constructed its own syllabus based on the Core Content. Your syllabus will include items in addition to those in the Core Content and you should get a copy of your syllabus. The contents of this book have been designed to cover the Core Content and the main additional sections required by each of the examination boards.

The purpose of a GCSE examination is to see what you:

Know, understand and can do

Examinations will require you to be able to demonstrate that you do understand, and that you can apply your knowledge and understanding. The feeling or flavour of GCSE will therefore be:

The appliance of science

1 Get a copy of your syllabus.

2 Compare your syllabus with the contents in the seven chapters. Sections of this book not included in your syllabus are not needed for revision purposes.

3 Make the decision to revise.

4 Decide how you will revise.

5 Plan a revision programme.

6 Stick to it.

7 Don't panic.

How do you revise?

There are many ways of revising. Unfortunately, most schools/colleges do not really prepare students in the art of revision. The following plan of attack may prove useful:

1 Work on your own at a desk or table.

2 Work in a quiet place, with perhaps the radio on very softly.

3 Avoid interruptions.

4 Make a revision programme.

5 Take regular breaks.

Here are some of the more productive methods of revising:

1 Read a section then:

    jot notes or key words down .... check yourself

    write out any formulae .... check yourself

    try a test question .... check yourself

    list important ideas in order .... check yourself

2 Study a diagram, then:
   sketch the diagram .... check yourself

   list any labels .... check yourself

   jot down notes of what it does/how it works .... check yourself

3 Pick some examination type questions and research the answers

4 Read the description of a laboratory experiment, then:
   note any safety precautions .... check yourself

   sketch the apparatus .... check yourself

   list everyday applications .... check yourself

Sample revision programme

Revising for an examination should not mean that everything else stops. It is essential to be organized and arrange your work and leisure to fit in together. Take time off to go to a disco or watch 'EastEnders', play football or puff on the clarinet. You may even have a part-time or Saturday job.

Identify times in the week when you can settle down to revise without interruption. Spread your revision throughout the week. Don't try and do it all in one go. Make your revision a regular thing and not something 'special'. Two half-hour sessions, with a short break in between, four or five times a week, gets a lot of revision done.

Once you've found the time to revise make a list of what you will revise. You may well have some subjects that need more revision than others. Give them more time. Pay particular attention to any subjects that you might need for a job, college or an apprenticeship.

If you start your revision early enough it's much easier. Cramming is hard work. You might think that Christmas of your examination year is early: Easter is late!

Aims of the chapter

By the end of this chapter you should be able to:

1 Recall that:
   (a) forces are measured in newtons with a newtonmeter;
   (b) forces act in pairs;
   (c) weight is a force;
   (d) pressure is measured in pascals (N/m^2);
   (e) the atmosphere has weight and exerts a pressure.

2 Distinguish:
   (a) between mass, force and pressure;
   (b) constant or uniform speed, acceleration and deceleration.

3 Describe experiments to:
   (a) show the moments at work on a balance;
   (b) illustrate Hooke's law;
   (c) find the centre of gravity of a thin card;
   (d) measure speed, acceleration and deceleration using a ticker-timer.

4 Demonstrate a practical knowledge of the applications and effects of forces in everday situations.

What are forces?

Forces are at work everywhere. They are much easier to identify than you might have thought. Anywhere you can see these things taking place, then forces are at work: attracting; bending; breaking; pulling; pushing; rotating; slowing down; squashing; speeding up; stopping; turning; twisting.

Forces are usually responsible for doing one of two things. They can **start things happening** or **stop things happening**.

For example, if you push a car or bike it will start to move. A force is at work. Forces are measured in **newtons** (symbol N) using a **newtonmeter**.

Forces in pairs

No matter where forces appear, they always appear in pairs. When you stand on the ground, the earth must be pushing up on you. If it didn't, you would sink. This is an example of an **action/reaction pair** of forces. The action of your force pushing down is opposed by the reaction force provided by the ground (in the opposite direction). Although action/reaction pairs of forces are very common, it is more usual in practice to work with single (action) forces only.

Changes made by forces – deformation

The word deformation means simply 'to change shape'. Forces that change an object's shape are very common. If a force is too large, it may change the shape of an object permanently or destroy it altogether. If the shape change is only temporary, then it is known as **elastic deformation**. A spring is a useful example of elastic deformation. When the stretching force is removed, the spring returns to its original shape. However, if the stretching force increases, it may stretch the spring beyond the limit of elastic deformation. Once the **elastic limit** has been passed, the spring will never return to its original shape no matter what is done to it! To make the best use of springs, the elastic limit should not be exceeded.

Robert Hooke investigated springs about 300 years ago and from this work comes **Hooke's law**:

> The extension of a spring is directly proportional to the applied force provided the elastic limit is not reached.

This is most easily understood by looking at a graph of load force against total extension (*not* total length) as shown in Fig. 1. The purpose of 'directly proportional' in Hooke's law is to describe a graph that increases evenly and starts at zero. In the example, the total extension doubles each time the load force doubles. When the load force is zero the total extension is zero.

The squeezing or compressing of an object will also cause a shape change. A road bridge will often have various heavy or light loads scattered across it. The bridge materials have to withstand these changing conditions. Walking creates

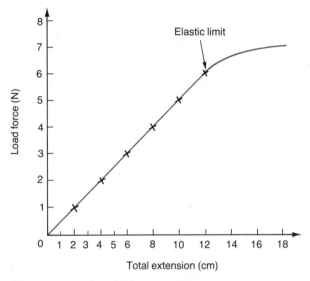

Fig. 1 A typical graph for a small spring

compressive forces on the various parts of the feet. One of the
ways in which feet and bridges cope with changes in compression
is to pass the forces on. Arched structures allow for the passing on
of forces to the supports (Fig. 2). Any loads appearing on a bridge
will be passed directly to the supports.

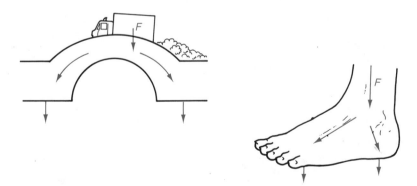

Fig. 2 Passing forces on

Changes made by forces – turning effects

A pair of forces applied to the ends of a pair of handlebars can keep the bicycle moving in a straight line or cause it to change direction. If both forces on the handlebars are working in the same direction then they will begin to turn. The turning takes place around the centre of the handlebars. This position is known as the **fulcrum**, or pivot. A force must be at work some way from a pivot if turning is to take place. A turning effect where the force acts at a distance from the pivot is known as a **moment**. The distance measured is always taken as the perpendicular (90°) distance from the pivot.

Turning moment = Force × Distance
(Nm)   = (N) ×   (m)

Turning moments are at work everywhere: door handles, screwdrivers, spades, crowbars, bicycle pedals, taps, the list is endless. The turning moments calculated in Fig. 3 have been chosen to be the same, although they are 'made up' in different

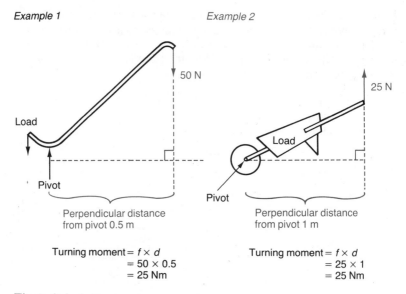

Example 1   *Example 2*

Load

50 N

25 N

Load

Pivot

Pivot

Perpendicular distance from pivot 0.5 m

Perpendicular distance from pivot 1 m

Turning moment = *f* × *d*
= 50 × 0.5
= 25 Nm

Turning moment = *f* × *d*
= 25 × 1
= 25 Nm

Fig. 3 Calculating turning moments

ways using different forces and distances. Each provides a moment of 25 Nm. The purpose of each turning moment is to 'move' something that opposes it. In the case of the crowbar, the opposition may be a crate lid. The load force in each example opposes the turning moment applied to the handles. In practice neither moment wins and the result is balance. The crowbar just lifts the crate lid and the wheelbarrow just lifts the grass cuttings. Both opposing moments must equal the moments applied to the handles.

A simple laboratory experiment to investigate opposing moments may be carried out in the following way (see Fig. 4). A metre rule is placed on a pivot in a position of balance. Different forces are then suspended from either side of the pivot. They will have to be at different positions to maintain a balance. This can be repeated and recorded a number of times. The moment on the left-hand side of the pivot is trying to turn in an anticlockwise direction and so is known as the **anticlockwise moment**. The moment on the right-hand side of the pivot is known as the **clockwise moment**.

For balance to be achieved:

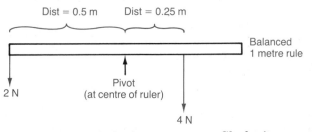

Anticlockwise moment = Clockwise moment

	Force (N)	Dist (m)	Moment (Nm)	Force (N)	Dist (m)	Moment (Nm)
1	2.0	0.5	1.0	4.0	0.25	1.0
2	2.0	0.4	0.8	4.0	0.2	0.8
3	5.0	0.25				
4	6.0	0.2				

Fig. 4 Moments at work

In the example of the crowbar (Fig. 3), assume that the perpendicular distance of the load force from the pivot is 0.1 m.

The clockwise moment is $(50\text{ N} \times 0.5\text{ m}) = 25\text{ Nm}$. The anti-clockwise moment must also be equal to 25 Nm.

$$
\begin{array}{llllll}
\text{Turning moment} & = & \text{Force} & \times & \text{Distance} \\
25 & = & ? & \times & 0.1 \\
25 & = & 250 & \times & 0.1
\end{array}
$$

The load force required for this to be correct is 250 N. This demonstrates the use of a lever as a force 'magnifier'. A force of 25 N applied to the handle has been magnified to 250 N at the crate lid. A screwdriver works on the same principle.

Changes made by forces – motion

To give a fuller picture of changes in movement brought about by the action of forces, this section will be divided into parts.

Simple ideas of speed

Speed is all to do with moving from place to place. Moving about takes time. Usually people talk of the distance moved in a single time period, often 1 hour. A long car journey may involve moving at a rate of 60 miles every hour. This is shortened to 60 mph. For each hour's travel, a further distance of 60 miles would be covered. In the following sections the measurement and discussion of speed will only use metres (or centimetres) and seconds, i.e., metres/second (m/s). (Any set of units for distance and time can be used as long as the same units are used throughout an experiment or question.)

In practice there is little difference between the words **speed** and **velocity**. They can usually be swapped for one another without difficulty. In Physics, velocity is used to describe speed in a particular direction.

Uniform or constant speed

If the speed of an object does not change from the beginning of the journey to the end of a journey, then it is said to be **constant** (or **uniform**). The total distance travelled depends on the speed of the journey and the time taken. The equation linking distance, speed and time is:

Distance = Speed $\times$ Time

For example, at 30 m/s (approx 70 mph) it will take 5 seconds to cover 150 m.

The other arrangements of the equation are:

$$\text{Speed} = \frac{\text{Distance}}{\text{Time}} \quad or \quad \text{Time} = \frac{\text{Distance}}{\text{Speed}}$$

Two types of graph are commonly used to illustrate the motion of an object. The graph in Fig. 5(a) uses information about speed and time. It is the **speed/time graph** (sometimes it is called a **velocity/time graph**). In this particular example, the speed value does not change as time increases. It represents a constant or uniform speed of 25 m/s. A very important feature of a speed/time graph is that the area under the graph line has the same value as the distance travelled in the journey.

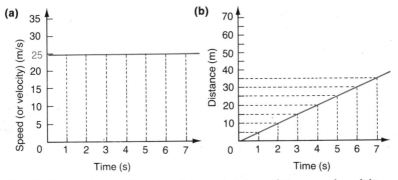

Fig. 5 Constant or uniform speed shown in (a) speed/time graph and (b) distance/time graph

The second sort of graph is the **distance/time graph** (see Fig. 5b). This example also shows a constant speed. During each extra second the distance away from the origin increases by a constant amount, in this case by 5 m every second.

Acceleration
An increase in speed (or velocity) is called **acceleration**. It is a change in speed that happens in one second. If a car accelerates

from 10 m/s to 15 m/s in one second, then the change in speed is 5 m/s. If the acceleration continues during the next second there will be a further increase in speed of 5 m/s, this time from 15 m/s to 20m/s. If the increase continues every second for a number of seconds the acceleration would be '5 m/s every second'. This is an example of constant acceleration. **The units of acceleration are m/s^2.**

The simple equation used to find values of acceleration is:

$$\text{Acceleration (m/s}^2) = \frac{\text{Increase in speed (or velocity) (m/s)}}{\text{Time taken for the increase in speed (s)}}$$

Graphs can easily be used to show information about accelerations. Fig. 6(a) shows a speed/time graph where the moving object is accelerating at 2 m/s^2. As each second passes the speed increases by 2 m/s. The area under the graph has the same value as the distance covered during the journey. The shape this time is a triangle not a rectangle. The curved line in Fig. 6(b) represents the basic shape of a distance/time graph for something travelling with a constant acceleration.

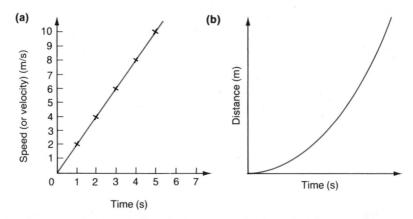

Fig. 6 Constant or uniform acceleration shown in (a) speed/time graph and (b) distance/time graph

Decelerations
Deceleration means slowing down. The same ideas, units and equations apply to deceleration as they do for acceleration.

Decelerations are **negative quantities** and should always be labelled with a **negative sign**. The graph in Fig. 7 shows thinking and braking time needed to stop an average car from 30 m/s (approx 70 mph).

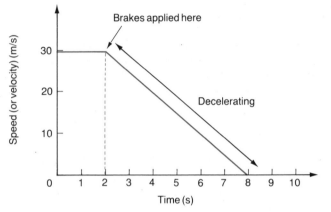

Fig. 7 Braking and thinking time needed to stop a car from 30 m/s (approx 70 mph)

Example Find the deceleration of the car in Fig. 7 in stopping from 30 m/s in 6 seconds. How far does the car travel during the two stages of the journey?

Deceleration = −Change in speed/time taken for change
                = −30/6   = −5 m/s^2
The negative sign indicates a deceleration.

The total distance travelled can most easily be found from the area under the graph. Divide this space into convenient shapes. In this case, a rectangle *and* a triangle.

Area of rectangle = 30 × 2 = 60 m
Area of triangle   = 30 × 6 × ½ = 90 m
Total   = 60 m + 90 m = 150 m

Forces make changes
Forces are at work with uniform speed, accelerations and decelerations. To keep anything moving with uniform velocity,

the **force of friction** has to be overcome. No matter how small that force is, it will eventually bring any moving object to a standstill. If an object is to be stopped quickly, then a braking force will have to be applied to that object. In that case, a force will cause a deceleration. The reverse is true too. To make an object speed up, a force will have to be applied to it. A practical example of this is the additional force required to make a bicycle accelerate.

 When a force is applied to an object, it will cause that object to either accelerate or decelerate. This is **Newton's second law of motion**. It is usually expressed in this form:

Force = Mass × Acceleration
(N) = (kg)   (m/s^2)

Example What force is required to increase the speed of a 2 kg mass by 3 m/s^2?

$F = m \times a$
$= 2 \times 3$
$= 6$ N

Had the example involved a deceleration, then the solution would have been a negative answer.

The use of ticker-timers
When connected to an alternating power supply, the ticker-timer causes a metal arm to vibrate 50 times every second. The metal arm has a raised spot on it. A piece of ticker-tape is drawn through the timer by a moving object. Each time the arm vibrates a dot is printed from carbon paper onto the piece of tape. Each gap represents the distance travelled in 0.02 seconds. Fifty gaps (51 spots) represents a time period of one second. In this way a piece of tape connected to a moving object passing through the ticker-timer produces a 'picture record' of the motion of that object, a series of dots (see Fig. 8(a)). Large gaps between pairs of dots will mean the tape is travelling fast. Small gaps between pairs of dots will mean a slow speed. Looking at the gap size at one part of the tape and comparing it with the gap size at a different part of the tape will tell you whether or not an object is slowing down or speeding up.

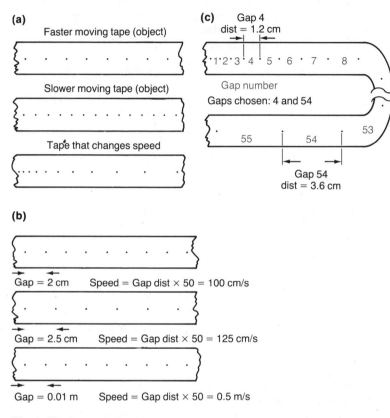

Fig. 8 Finding accelerations

The first example shown in Fig. 8(b) has a gap spacing of 2 cm. As 50 gaps pass by, the total distance travelled in one second is 2 cm × 50 = 100 cm. The constant speed of that tape is 100 cm/s.

Speed = Gap distance × 50

Accelerations and decelerations are most easily found with a ticker-timer if the tape has more than 50 gaps on it. Number all the gaps on the tape. Choose a pair of gaps 50 spaces apart. (By doing this you have chosen two spaces with an interval of one second.) Using the method above, find the speed of the two sections of tape (1 second apart). The difference between these two values is the change in speed during a one-second interval. This is the definition of acceleration/deceleration, and gives you the required value straight away.

Example   Find the acceleration recorded on the tape in Fig. 8(c).

Gap 4 distance = 1.2 cm
            Speed   = 1.2 × 50
                    = 60 cm/s

Gap 54 distance = 3.6 cm
            Speed   = 3.6 × 50
                    = 180 cm/s

Difference in speed in one second = 180 − 60 = 120 cm/s

Measuring forces

Two of the most convenient ways of measuring a force are by stretching a spring or bending a beam. The **newtonmeter** or spring balance depends on **Hooke's law** to provide a robust and fairly reliable method for measuring force. Often they are marked in kilograms and newtons.

    In practice when a thin (springy) beam is held horizontal it will begin to bend. It will bend further if a force is applied to the far end. The greater the force the greater the bending. By placing a scale at the end, the beam can be calibrated into a device for measuring force.

Weight is a special force

The action of gravity causes an object to be pulled towards the earth. Every object feels this force pulling down on it. The force acting downwards is given a special name: **weight**. Weight is a force and so it is measured in newtons. An object can only have weight if gravity is at work. If there is no gravity there can be no force so there can be no weight. Every object has what is known as **mass**, the material it's made from. The mass of an object cannot change, no matter where you put it. What can change is the force that pulls on an object. In space there is no gravity, so no force can pull on it. An object will have mass (material) in space, but no weight. On the moon the force due to gravity is less than it is here on earth. The mass will be the same, but the weight will be less.

    Weight is calculated from the equation given for Newton's second law:

Force = Mass × Acceleration

By changing the wording (but not the meaning) very slightly:

Force on the ground = Mass × Acceleration due to gravity
or   Weight (W) = Mass (m) × Acceleration due to gravity (g)
$$W = m \times g$$

The usually accepted value for g is about 10 m/s^2. A 1 kg mass will have a force pushing down on the ground equal to 10 N.

Stability and centres of gravity

The **centre of gravity** of an object is the position at which the weight of that object seems to be concentrated. It is the position at which gravity seems to act.

To find the centre of gravity (C of G) of a thin piece of card is easy to organize. Pin the edge of the card to a notice board. It must be free and easy to swing. Hang a plumbline from the same pin. The C of G of the card will always hang directly under the pivot (the pin). With a pen, mark the position of the plumbline on the card. The C of G lies along that line. Repeat the process twice more from different places along the edge. The C of G of the card is where the three plumbline marks cross.

The ability of an object to stay in its original position is known as **stability**. Any object will remain in its original position unless the C of G moves to a new position outside its base area. There are three distinctly different examples of stability. A full bottle of milk in the upright position would be **stable**; it would be **unstable** standing on its neck, and in a **neutral** position on its side.

Pressure

Pressure is the spreading out of a force onto a surface. A large force concentrated onto a small area would exert a higher pressure than a small force spread out over a large area. Pressure is the force placed on a single unit of area. The area chosen is usually 1 m^2 or 1 cm^2. Pressure allows for the transmission of forces from one place to another. In solids, pressure acts only in the direction of the force. In fluids (liquids and gases) **pressure acts in all directions**.

$$\text{Pressure} = \frac{\text{Force}}{\text{Area}}$$

The SI units are N/m^2; $1 \text{ N/m}^2 = 1$ **pascal** (Pa). It is often more practical to deal in cm^2; the units of pressure then become N/cm^2.

Pressure and solids
The drawing pin is often used as an example of pressure transmission in solids (Fig. 9). Whatever force is applied to the head of the drawing pin must be passed onto the point. At the head of the pin, the force is spread over a large area. The pressure here is small. At the point that same force is spread over a very tiny area. The pressure here is very large indeed. The design of the pin concentrates a force at the head into a much smaller space at the point. As a result of this, the point will pierce the surface of the material.

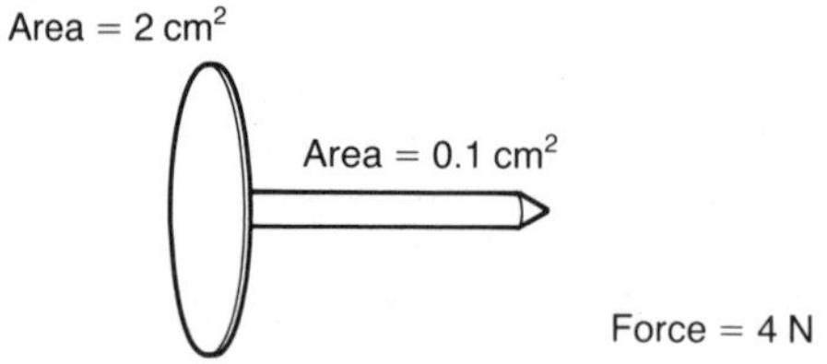

Area = 2 cm²

Area = 0.1 cm²

Force = 4 N

Pressure on head = $\dfrac{4\,N}{2\,cm^2}$ Pressure on point = $\dfrac{4\,N}{0.1\,cm^2}$

= 2 N/cm²   = 40 N/cm²

Fig. 9

Pressure in fluids
Fluids behave in a different way from solids. They can flow in any direction. If a force is applied to a fluid it will be passed on in all directions. It will be transmitted through the fluid to the walls of the container. For example, if a water butt fills with rain the weight of the water provides the force, which is then passed to the walls of the container.

Example A water butt fills with 20 kg of water. If the area of the base of the butt is 0.25 m² calculate the pressure on the bottom surface.

20 kg provides a downwards force (the weight) of 200 N.

$$\text{Pressure} = \frac{200\,N}{0.25\,m^2} = 800\,N/m^2$$

The water exerts a pressure equal to 800 N on every square metre.

In this example if the force were greater, the pressure would be greater too. To increase the force, more water would have to be added to the butt. **Pressure increases as the depth of the fluid increases**. This is true for any fluid.

The pressure of the atmosphere decreases as the height above the earth's surface increases. Aeroplanes have to be 'pressurized' when they fly at high altitudes. The atmosphere can be thought of as a column of air approximately 10 km high. It has weight and provides a downward force. A column of water approximately 10 m high has the same weight; so too, does a column of mercury about 76 cm high.

A **barometer** is used to measure atmospheric pressure. Two forces are at work: the weight of the atmosphere; and the weight of the mercury column. The pressure of the mercury column is found from its height (Fig. 10a).

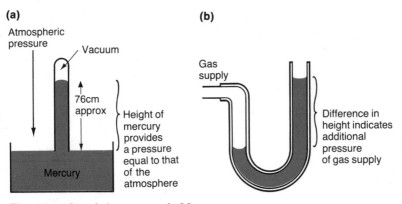

Fig. 10 (a) Simple barometer (b) Manometer

A **manometer** works on the same principle as a barometer, but uses water instead of mercury. Manometers are used to measure excess pressures in gas pipes. A 'U' shaped length of tubing is partially filled with water. One end is connected to a gas supply. The gas pressure will cause the water in one arm of the tubing to rise. The difference between the two water levels is used to indicate the gas pressure (Fig. 10b).

Hydraulic systems

The ability of fluids to pass on pressure in all directions has many uses in industry and engineering. By using a system of piping it is possible to pass energy into often difficult and remote places. A car braking system is one example of using pressure transmitted by fluids (Fig. 11).

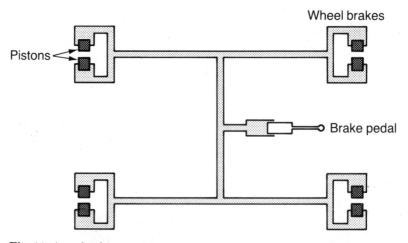

Fig. 11 A car braking system

Summary

Forces cause changes to take place. Forces are measured in newtons and always appear in pairs. When a force is spread out over an area it is known as pressure. In fluids pressure increases with depth and acts in all directions. Weight is the force due to gravity. Friction always opposes motion. Hooke's law links force with extension for an object being stretched, e.g., a spring. Speed, acceleration and deceleration can be 'pictured' on ticker-tape.

Practice questions

Try these questions first without reference to the text. Check your answers with the specimen answers at the back of the book. Answers to questions involving calculations should show all the working and should contain a small amount of explanation, including any equations used.

1 In the example of the wheelbarrow (Fig. 3), what load would be suitable if it were placed 0.4 m from the wheel centre? The turning moment at the handle remains the same at 25 Nm.
2 Fig. 4 gives some completed examples for calculating turning moments. Complete the calculations on the left-hand side of the table and make some suitable suggestions for values on the right-hand side.
3 Use the speed/time graph in Fig. 6 to find the distance travelled during the first 5 seconds of the journey. If the speed of the object then levelled off, how much farther would it travel in the next 10 seconds?

2 Energy

Aims of the chapter

By the end of this chapter you should be able to:

1 Recall that:
   (a) energy makes things happen;
   (b) all forms of energy are measured in joules;
   (c) energy can be changed from one form to another.

2 Distinguish between:
   (a) the various energy forms;
   (b) energy and power;
   (c) heat and temperature.

3 Describe experiments to:
   (a) show methods of energy transfer;
   (b) find the fixed points of a thermometer;
   (c) test for the absorption and emission of heat energy.

4 Demonstrate a practical knowledge of energy sources and methods of energy conservation, thermometers and temperature control devices.

Do not be tempted to slip through the first part of this chapter. This is what Physics is all about. Physics is the study of energy and energy changes.

What is energy?

There is no simple answer to the question 'What is energy?' A much simpler question to answer is 'What can energy do?' If something has energy it can do something – it's as simple as that. The things in this list seem to have nothing in common: a chocolate bar; a battery; kicking a football; a 'sun' lamp; a plant; the moon. But they are all connected with the idea of energy. They can all do something or cause something to happen.

Energy forms

Rather than have a never-ending list of items that have energy, it is more convenient to arrange them into sets or groups. For example, all things that move have energy. Anything that moves is said to have **kinetic energy** (KE). In a similar way, anything that stores energy for use at a later time is said to have **potential energy** (PE). The energy forms that are most commonly used in discussions of Physics are (see Fig. 1):

Chemical energy Heat energy Nuclear energy
Potential energy Magnetic energy Sound energy
Kinetic energy Electrical energy Electromagnetic wave energy

Sometimes PE and KE are considered together under the heading **mechanical energy**, and sound and electromagnetic waves are considered under the heading **wave energy**.

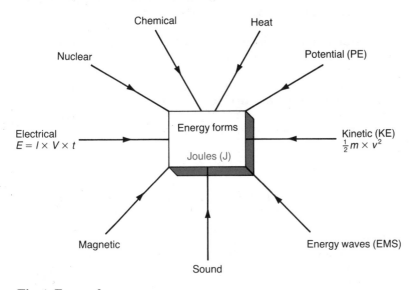

Fig. 1 Energy forms

Each of these energy forms will now be dealt with in greater detail. Do not jump to the idea that the list is incomplete and that energy forms are missing. Items such as wind and water have been left out because, by themselves, they are not energy forms.

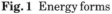 Chemical energy

This form of energy is not always as obvious as the name might suggest. Chemicals such as oil, coal and gas would be included in this list. They can all be burnt to provide heat. The chemicals inside a battery will react together to provide electricity. It might be more correct to say that chemicals are **stores of energy** that can be released at any convenient time. Perhaps chemicals belong to another energy form – potential energy? Chemical energy is measured in joules (J).

▶ K Potential energy (PE)

A battery is an example of **stored energy**. When it's being used it changes the chemicals inside to produce electricity. Water really has no energy by itself, but if you put it high up behind a dam wall and allow it to fall down and drive a turbine, then it has energy. It only has energy because of its position. Any object at ground level really has no PE, but when it's raised from the floor it has. Anything that is high up will have the 'potential' to be pulled down by gravity. If an object has PE because of its position, it may sometimes be called 'gravitational energy'. No matter how PE is stored, it is measured in joules (J).

▶ K Kinetic energy (KE)

Any object that moves has some KE. People walking have KE, a kicked football has KE. Water at the top of a waterfall has PE, but as soon as it drops over the edge it begins to lose height and gain speed. It loses PE and gains KE. KE is measured in joules (J). The amount of KE a moving object has increases with mass (kg) and velocity (m/s). The equation used to calculate the KE of a moving object is:

$$\text{Kinetic energy} = \tfrac{1}{2} \times \text{Mass} \times \text{Velocity}^2$$
$$\text{(J)} \qquad \text{(kg)} \qquad \text{(m/s)}^2$$
$$\text{KE} = \tfrac{1}{2} \times m \times v^2$$
$$\text{KE} = \tfrac{1}{2}mv^2$$

Example Find the energy of a car, mass 1000 kg, moving at a speed of 30 m/s (approx 70 mph).

$$\text{KE} = \tfrac{1}{2}mv^2$$
$$\text{KE} = \tfrac{1}{2} \times 1000 \times 30 \times 30$$
$$\text{KE} = 500 \times 30 \times 30 = 450\,000 \text{ J}$$

Total KE of the car is 450 000 J (450 kJ).

Heat energy

Heat is an energy form. Heat can make things happen. If heat is added to ice, the ice melts. When more heat is added, the water eventually turns to steam. An iron bar will expand when heated. Every material has some heat energy inside it. Ice may feel cold, but it still contains some heat energy, even though the amount may be small. Heat energy is measured in joules (J).

Magnetic energy

It may not seem likely at first, but magnetism is a form of energy. It is probably the least important energy form on the list as far as GCSE Physics is concerned. Magnetism can make things happen. It will attract other materials to it and make them move. As the energy is stored in a magnet, perhaps magnetism is really a form of PE. Magnetic energy is measured in joules (J).

Electrical energy

It is almost impossible to hold on to electricity. A battery does not store electricity, it stores chemicals that provides electricity when the battery is connected into a circuit. Electricity can certainly make things happen. It is the most versatile energy form. It is the most **easily converted** energy form. For example, a bulb will give out light (and heat) when electricity passes through it. The amount of electrical energy passing through a device depends on: the current value in amperes; the potential difference in volts; and the time, in seconds, for which the circuit is switched on.

Increase any of these and the total energy will increase. Electrical energy is measured in joules (J). The equation used to find electrical energy is:

$$\begin{aligned}
\text{Energy} &= \text{Current} \times \text{Potential difference} \times \text{Time} \\
\text{(J)} &= \text{(amperes)} \times \text{(volts)} \times \text{(seconds)} \\
E &= I \times V \times t \\
E &= IVt
\end{aligned}$$

Nuclear energy

This form only needs to be mentioned briefly here. As atomic nuclei break up they have a heating effect on their surroundings. Nuclear energy (in a nuclear reactor, for example) appears as heat. The heat is then used to do something useful, such as provide steam to drive a turbine (see Chapter 3, Matter). Nuclear energy is measured in joules (J).

Sound energy

Sound is an energy form. It is an energy form because sound is the **movement of molecules**. To be more precise it is the vibration of molecules. Sound energy is passed on from one molecule to another. As sound involves movement, should it be mentioned under the heading 'kinetic energy'? Sound can cause things to happen, it can cause windows to break or people to go deaf! **Do not** link the phrase 'sound waves' with ideas such as 'light waves' – they may seem similar, but they are really very, very different from one another. Because sound involves molecules, sound energy cannot exist in space. Sound energy is measured in joules (J).

Electromagnetic waves (EMS)

Any wave form such as light, X-rays, ultraviolet or gamma waves, belongs to a set known as the 'electromagnetic spectrum' (EMS). They are sometimes called **waves**, **rays** or even **radiations**. These titles are sometimes even mixed together, e.g. 'microwave radiation'. No one can really tell you what these energy waves are. It is sufficient to realize what they can do. For example, light waves will help plants to grow and will allow 'solar powered' calculators, etc. to work. Members of the EMS will cause materials on which they land to warm up. Some are better at this than others. Infrared radiation is extremely good at it. This particular radiation is responsible for the warming effect of a sunny day. Try to avoid phrases such as 'heat wave' when you really mean infrared radiation.

Other energy forms

Obviously there are some energy forms missing from this rather short list – or are there? Wind power has not been mentioned. The wind only has energy because it moves. It has KE. If you think an energy form is missing, it is more than likely to be 'hidden' under one of the previous headings.

Energy sources

It is important to be able to discuss energy forms as either
renewable or **non-renewable**. Coal and oil are plentiful today.
They were laid down in the earth's surface millions of years ago
but will not last forever. Energy sources such as these are finite
and non-renewable. Solar energy is plentiful, continuous and
unlikely to run out. It is important to be aware of alternative
energy sources. This list is not complete: solar power; wind power;
wave/tidal power; geothermal (such as the hot water springs in
Iceland, where water bubbles to the surface from underground);
recycling waste (such as burning rubbish to provide heat).

You should be able to discuss points for and against these
energy sources, both renewable and non-renewable, e.g.:

Points for	Points against
Wind power	
1 Free	1 Wind unreliable
2 Never-ending	2 Large aero-generators (windmills) would spoil the landscape
Solar Power	
1 Free	1 Unreliable weather in UK
2 Never-ending	2 No use at night
Nuclear power	
1 Reasonably cheap	1 Storage of waste
2 Enormous reserves	2 Danger of radiation fallout

Energy changes

Physics is all about energy and more importantly energy
changes. A very simple example can be used to illustrate this
idea. There are several major energy changes that take place
when a simple pocket torch is switched on.

(a)

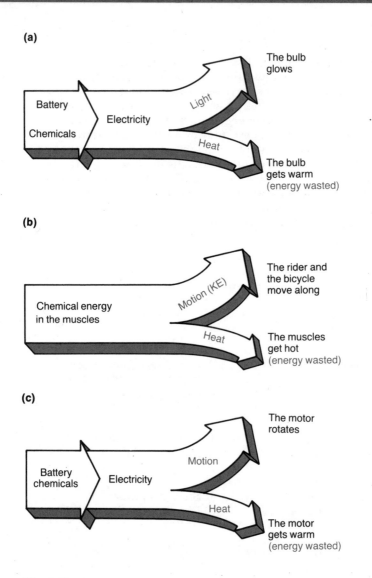

(b)

(c)

Fig. 2 Energy arrows

　Fig. 2(a) is an **energy arrow** that shows the major energy changes that take place. Fig. 2(b) is a similar energy arrow, but this time it is for a cyclist, and Fig. 2(c) is an energy arrow for a

battery driving an electric motor. It is often easier to identify an energy arrow than to spot an individual energy source. In all these examples some of the initial energy is wasted as heat. It is quite usual for energy, in the form of heat, to be wasted when one form of energy is changed into another. Energy arrows can be made up for any situation where energy changes take place.

Energy changers

A device that changes one form of energy into another is given the grand name: *transducer*. Table 1 gives some examples of transducers and the energy forms they link together.

Transducer (energy changer)	Main energy change	
	From	To
Battery	Chemical	Electrical
LED (Light Emitting Diode)	Electrical	Energy waves (light)
Oil	Chemical	Heat/light
Loudspeaker	Electrical	Sound
Microphone	Sound	Electrical
Battery charger	Electrical	Chemical
TV	Electrical	Heat/sound/light

Table 1 Some transducers

Fig. 3 overleaf shows some of the main energy changes that take place in the different types of power station.

Work

The word 'work' in Physics has a rather different meaning to that in use every day. The strict definition of **work**, as used in Physics, **requires a force to move**. Pushing a car to 'bump-start' it involves work. Holding a heavy bag of groceries does not. **Work is energy being used up.**

Work = Force × Distance
(J) = (N) × (m)

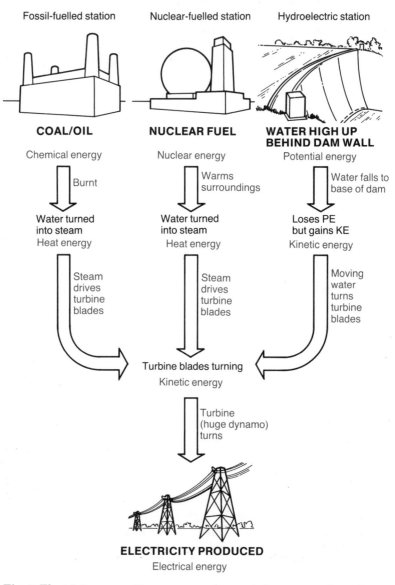

Fig. 3 Electricity generation: energy and energy changes are shown in colour

Working is a method of transferring energy. A simple example of this could involve picking a book up from the floor. By lifting it up you have given the book PE. There has been a net energy transfer from you to the book. Pushing a crate along the floor involves a transfer of energy from you to the crate to overcome friction.

In Fig. 4 the ideas of force and distance are quite obvious. The crate will require a force of 100 N to make it move. By moving it 20m, the amount of work done (i.e. energy used up) on the crate is:

Work done (J) = Force (N) × distance (m)
= 100 × 20
= 2000 J

You would have to supply 2000 joules of energy (from food?) to the crate to make it move.

Frictional force to push against = 100N

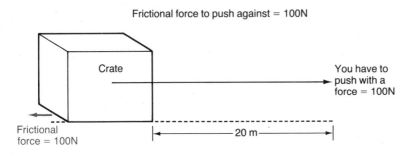

Crate

You have to push with a force = 100N

Frictional force = 100N

20 m

Fig. 4 Work done moving a crate

Example What is the work done in 'bump-starting' a car if you have to push it with a force of 700 N for a distance of 30 m? How much energy will you have to supply?

Work done = Force (N) × distance (m)
= 700 × 30
= 21 000 J

The work done on the car is 21 000 J (or 21 kJ). The energy supplied must be equal to the work done = 21 000 J (or 21 kJ).

Energy efficiency

The idea of wasted energy is an important one. The three diagrams in Fig. 2 show that some *energy is wasted* in the form of heat. Virtually all machines (including biological machines!) waste energy as heat.

The energy arrows in Fig. 5 show two different versions of the same machine. Marked on each machine is the energy input, the energy lost as sound, the energy wasted as heat and the useful energy extracted.

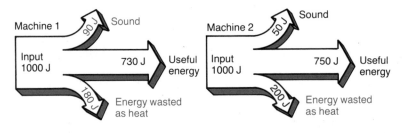

Machine No.	Total energy wasted	Most efficient machine
1	90 + 180 = 270 J	2nd
2	50 + 200 = 250 J	1st

Fig. 5 Finding the efficiency of a machine

Machine 2 is the most **efficient** because it wastes the least amount of energy.

Sometimes it is more convenient to present the efficiency of a machine, not just by how much energy it has wasted, but as a percentage.

The efficiency of any machine is found using the following equation:

$$\text{Efficiency} = \frac{\text{Energy output}}{\text{Energy input}} \times 100\%$$

or

$$\text{Efficiency} = \frac{\text{Useful energy extracted from a machine}}{\text{Total energy put into a machine}} \times 100\%$$

In the case of machine 2 in Fig. 5, the efficiency is:

Energy extracted from the machine 750 J
Energy put into the machine   1000 J

$$\text{Efficiency} = \frac{750\,\text{J}}{1000\,\text{J}} \times 100\% = 75\%$$

Energy conservation

When performing calculations always check your answer for common sense. A machine can never have an efficiency value greater than 100 per cent. (Even values close to 100 per cent are rare). The total energy output of a machine must always equal the energy input, but some will always be wasted in a form that's not very useful. More often than not wasted energy appears as heat/sound.

Energy and power are not the same

Do not confuse the words *energy* and *power*. Do not use one of them when you really mean the other. This is another example of words having special meanings in Physics. The word 'energy' has no connection with time; 'power' does.

To have energy means to be able to do something. An amount of energy is measured in joules (J).

To be powerful means to be able to use a large amount of energy all the time. **Power** is the *rate* at which (*how quickly*) **energy is used or transformed**.

For a simple example, imagine two identical batteries containing an identical amount of energy, say 20 000 J.

Battery	Total energy available	Amount of energy used each second	Time taken before battery runs out
1	20 000 J	100 J	200 s
2	20 000 J	2000 J	10 s

Battery 2 will run out of energy faster because it uses it up more quickly. This is the more powerful. Battery 1 is the less powerful because it uses less energy every second.

Power = Rate of energy consumption

$$\text{Power} = \frac{\text{Energy (J)}}{\text{Time (s)}}$$

The units of power are joules/second (J/s):

1 watt = 1 J/s

A machine with a power rating of 1000 watts will consume energy at the rate of 1000 J/s.
Power is measured in **watts (W)** (1000 W = 1 kW).

The final energy form: heat

Wherever and in whatever form energy starts out, it eventually ends up being *absorbed by molecules*. When a molecule takes in energy then it vibrates a little more violently. Its internal energy will increase. Its internal KE will increase. No matter how many energy changes take place, the energy form at the 'end of the line' is heat.

Heat and temperature

Do not confuse these two ideas. Quite often it is easy to use one of them when the other is required. Heat and temperature are closely connected, but they are not the same. Heat is an energy form. **Temperature** is a measure of *hotness* or *coldness*. It might prove useful to use the phrase 'heat energy' rather than just 'heat' to distinguish between heat (energy) and temperature (hot/cold). The comparison between a spark and a filled hot water bottle is often used to distinguish between the two ideas.

A spark is extremely hot. It has a very high temperature. It is small and contains little heat energy. Its heat energy is released almost instantly on contact.
A filled hot water bottle is only warm. It has quite a low temperature. It is large and contains a considerable amount

of heat energy. It will continue to release this energy for a
long time.

Heat energy transfer

There are three methods of heat energy transfer: convection;
conduction; and radiation. (For more detailed information about
molecules and their behaviour, see Chapter 3, Matter.)

Convection and convection currents

For convection of heat energy to take place the material
containing the heat must be a fluid (liquid or gas). The heat
energy is carried/transferred by the molecules in the material. It
is carried/transferred because the molecules move. They take the
heat energy from place to place. As they move they take the
energy with them. A movement of molecules is known as a
convection current. There are several simple laboratory
experiments available to demonstrate convection currents.
Fig. 6 shows how to demonstrate convection currents in liquids.
Fig. 7 shows how to demonstrate convection currents in gases.
Both work because of the idea that applies to most materials –
heating causes expansion (see Chapter 3).

In both examples the fluid containing more heat energy rises
out of the way, allowing less energetic material to move in behind
it. Energy is transferred in this way. Eventually all the fluid will
heat up.

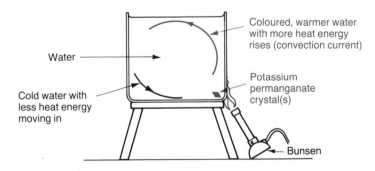

Fig. 6 Convection currents in liquids

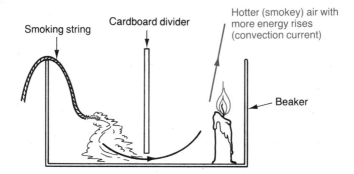

Fig. 7 Convection currents in gases

Kettles
A very practical use of convection currents can be found in a household kettle (see Fig. 8). The water is heated by the 'immersion heater' at the bottom of the kettle. The hot water convection currents rise to the top of the tank and cold water moves in to replace it. This continues until the water eventually boils.

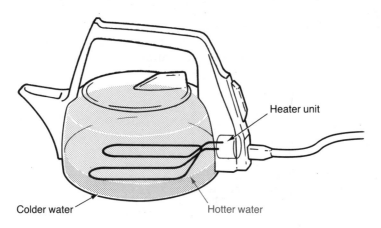

Fig. 8 Convection in a kettle

Sea breezes
Convection currents are responsible for sea breezes.

During the day the land heats up more quickly than the sea. Hot air rises over the land and is replaced by cooler air coming in from above the sea. A breeze is the result, blowing onshore from the sea (Fig. 9).

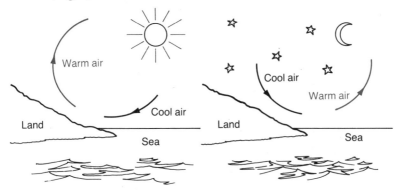

Fig. 9 Convection currents responsible for sea breezes

At night the situation is reversed. The land cools much more quickly than the sea. Warm air currents rise over the sea to be replaced by colder air from above the land. A breeze is the result, blowing offshore from the land.

The breeze(s) may die down for a short time around the time of sunset. (Why?)

Conduction

Conduction is the transfer of heat energy from one piece of material to another. A good **conductor** of heat will easily pass heat energy on. A bad conductor, called an **insulator**, will not pass on heat energy very easily. The main difference between conduction and convection is the movement of molecules.

Conduction takes place between molecules that are next door to one another. They do not need to move position to pass on heat energy.

Conduction in solids

Conduction can take place in solids. (Convection can only take place in fluids.) It can take place in fluids, but this is of less importance. Two methods of demonstrating the conduction of heat energy in solids are shown in Fig. 10. These are quite crude

but effective demonstrations. The results will give a list of
materials in order of 'best conductor', but no more than that. Each
experiment requires the ends of a set of rods to be heated. The
heat energy is conducted along each rod and eventually causes
the wax to melt, releasing the small lead shot. To make the
results more meaningful you could: ensure rods are equal
thicknesses; measure how long the heat takes to travel a
measured distance.

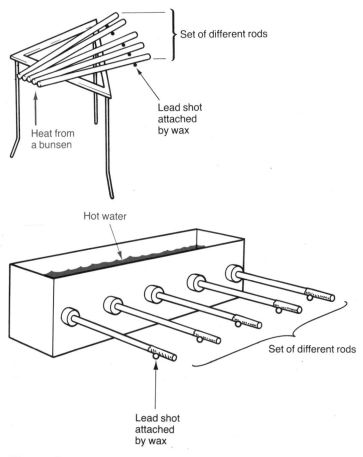

Fig. 10 Comparing conduction

Conduction in water
Water is a bad conductor of heat energy. It is very good at
convection! Because water will move, a little 'trick' is required to
show its conduction properties and stop convection currents
spoiling the experiment. A piece of ice is trapped at the bottom of
a test tube (see Fig. 11). The water at the top of the test tube is
heated to boiling. If water is a good conductor, then heat energy
will be conducted down through the main body of water to the ice,
causing it to melt quickly. This does not happen.

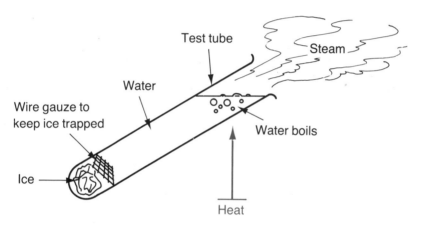

Fig. 11 Water is a bad conductor of heat energy

The rate of conduction
The rate at which heat energy is conducted through a material
depends on several things:

1 The thickness of the material A thicker material would be able
to transfer more energy than a thinner material.
2 The temperature difference A higher temperature difference
between the two ends of a piece of material (temperature
gradient) will mean a higher rate of energy transfer.
3 The material itself Different materials conduct heat energy at
different rates. For example, copper will transfer heat energy
much faster than an identical piece of glass.

Heat energy transfer: radiation

There are many different kinds of radiation. The particular variety of radiation linked with heating is **infrared radiation**. These radiations are variously called 'infrared radiations', 'infrared rays' or 'infrared waves'. All three are identical. Radiations are emitted or given out by a source. Anything that is hot and contains heat energy is a potential source of infrared radiation. A hand will feel hotter when placed in front of an electric fire. The fire **radiates** infrared radiation *directly*, in all directions, to whatever is nearby.

Detecting infrared radiation

When infrared radiation falls on to a **phototransistor** connected to a milliammeter it triggers a current to be switched on. This registers on the meter. A **thermopile** produces an electric current directly. This, too, can be observed with the use of a sensitive ammeter. A **thermistor** can be used instead of the phototransistor. These devices are shown in Fig. 12.

Radiation is direct. It travels in straight lines. The sun is very hot and emits enormous amounts of infrared radiation. Unlike the other methods of heat transfer, infrared radiation can *travel through a vacuum*. Convection and conduction require a material for their method of heat transfer to work.

Absorbing and emitting infrared radiation

Some surfaces are good absorbers of infrared radiation and some are bad. Some surfaces are good at emitting infrared radiation and some are bad.

Testing absorption

Testing materials to see how well they absorb infrared radiation can easily be done using the apparatus shown in Fig. 13. It is a set of four plates with different surfaces and textures. A drawing pin has been stuck on the back of each plate with wax. All four plates are fixed the same distance from the heater. When the heater is switched on the four plates will start to absorb the heat energy. One by one the drawing pins will fall off. The drawing pin will fall off the back of the best absorber first, and so on.

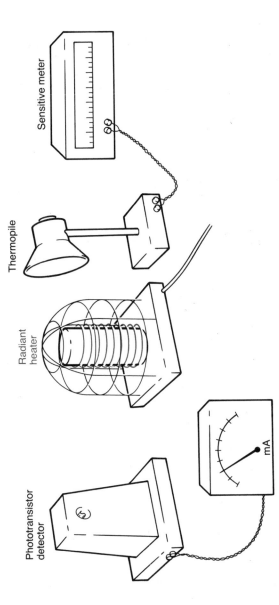

Fig. 12 Two ways of measuring radiant heat

Fig. 13 Find the best absorber

A second experiment uses a 'black' flask and a 'silver' flask, each containing a thermometer. They are both placed the same distance from a heater. There will be a different rise in temperature on the thermometer in the 'black' flask when compared to the temperature rise in the 'silvered' flask.

Dark, dull colours are the best absorbers.
Bright, shiny colours are the worst absorbers (which means they are the best *reflectors*).

A black car is noticeably hotter inside on a bright sunny day, than the same car in white.

Testing emission

'Leslie's cube' is a convenient piece of apparatus to test surfaces for their ability to emit infrared radiation. It is a copper cube. Two sides are painted; one black, the other white. The third side will be dull copper and the fourth, shiny copper. The cube is filled with hot water and the thermopile is used to measure the heat given off from each side in turn. The greater the heat energy emitted the larger will be the current given out by the thermopile and displayed on the meter.

Dark, dull colours are the best emitters.
Bright, shiny colours are the worst emitters.

Insulators

It is wrong to think that insulators keep the cold out. It is not possible to transfer 'cold' to anywhere; *only heat can move.* Insulators keep the heat in, not the cold out. An insulator is a barrier to heat flow.

The Thermos flask

This household utensil is known commonly as a *Thermos*, a *vacuum flask* or a *Dewar flask*.

The flask was designed to stop the transfer of heat. It is equally effective at keeping things hot (by stopping heat escaping) or keeping things cold (by stopping heat entering).

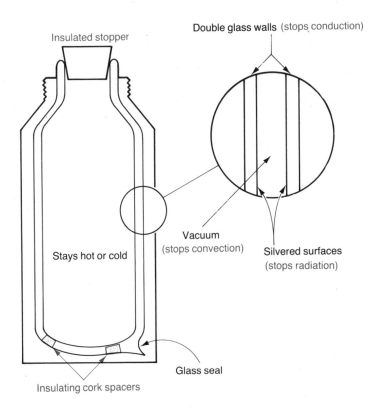

Fig. 14 The common vacuum flask

The flask is a particularly useful demonstration device because its construction is designed to *reduce the heat loss due to all three methods of heat transfer* (see Fig. 14).

The flask is made of glass which is a *good insulator*. This prevents heat loss by conduction. Heat loss by convection is due to molecules moving, so taking the heat with them. A perfect vacuum would eliminate this form of heat loss completely. In practice, this vacuum is not perfect. The *silvered surfaces* of the flask are included to act as mirrors. Their job is to reflect the infrared radiation back in the direction from which it came.

Testing the vacuum

Use two identical flasks, one with the glass seal intact and one with the glass seal broken. The flask with the broken seal will have an air layer between the glass walls. Fill each flask with an identical amount of hot water (at the same temperature). Use stoppers with thermometers in them to observe the temperatures of the flasks over a period of time. Plotting a **temperature/time graph** would be a useful exercise. You would expect to find the flask with the broken glass seal to cool more quickly. A second experiment might be to test different flasks (with their seals intact) to compare their insulating properties. Measure their temperatures over a period of time and plot a cooling curve for each flask. A typical graph (using two flasks) is shown in Fig. 15.

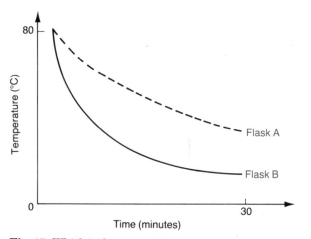

Fig. 15 Which is the more efficient insulator – flask A or flask B?

Air as an insulator

Air can be a good insulator, but only if it is stopped from moving. When convection currents start flowing, air transfers heat energy very well. Any system that traps pockets of air will be an effective insulator. The following is a list of practical examples of *trapped air insulators*: carpets; blankets; string vests (only when under something else, e.g., a shirt); bird feathers; duvets; double glazing; fur/hair; a kilt; cardboard/paper. When answering questions about items such as those listed, include information about 'trapped air as an insulator'.

Household insulation

You should be aware of effective household insulations including: carpets; draught excluders; double glazing; roofing insulation; cavity wall insulation; reflective boards behind central heating radiators; lagging of pipework and water tanks.

The *U-value* ratings shown on some insulating materials refer to their ability or lack of ability to insulate against heat loss. The lower the U-value the better the insulating properties of the material.

Household double glazing

Two sheets of glass trap a layer of insulating air. The air gap between the glass panes is the important factor. If the gap is too large, convection currents will begin and the insulating properties will lessen.

Temperature and temperature control

The measure of temperature is the measure of hotness and coldness. There are two scales in common use, the *Celsius scale* and the *Kelvin scale*.

The Celsius scale is wrongly called 'centigrade'. It has this nickname because the scale of temperature is divided into 100 divisions. On the Celsius scale (°C) 0 °C represents the temperature at which pure water freezes. The 100 °C mark represents the boiling point of pure water.

A second temperature scale is also now in common use. The *Kelvin* (K) or 'Absolute' scale starts at a temperature much lower than that of the Celsius scale. The zero mark (0K) on the Kelvin scale is the temperature at which *molecules cease to move*.

To convert °C to K   *add* 273
To convert K to °C *subtract* 273

The mercury-in-glass thermometer

Mercury is used because of its rapid reaction to temperature change. It is also easy to see. Different varieties can be manufactured to cope with temperature changes in the range −10 to 300°C. The walls containing the mercury reservoir are *very thin* to *allow fast conduction* of heat. A narrow bore tube also contributes to the sensitivity of the thermometer. Mercury freezes at −39°C so is no good for measuring low temperatures.

The clinical thermometer

This is basically a mercury-in-glass thermometer, but with two important changes (see Fig. 16(a)):

1 A 'kink' in the narrow bore tube prevents the mercury thread from backing into the reservoir, by causing it to break. Shaking the thermometer restores the mercury to the reservoir.
2 The thermometer only has a very small temperature range, usually 35 to 42°C. The normal human body temperature of 37°C is usually marked.

The alcohol thermometer

Alcohol boils at a fairly low temperature (78 °C) but freezes at −112°C so it is ideally suited for the measurement of very low temperatures.

The thermocouple

When a pair of different wires, such as copper and constantan or iron, are brought together and heated (or cooled) they generate a tiny electric current. A suitable arrangement is shown in Fig. 16(b). Thermocouples can be manufactured to measure temperatures in the range −200°C to above 1800 °C. They can be used in places such as furnaces or the core of a nuclear reactor.

Thermistors

The resistance of *semiconductor devices* to an electric current decreases with temperature. The circuit diagram in Fig. 16(c) could be used as a fire alarm. Path 1 will not work until the thyristor is switched on by an electric current travelling in path 2. Not enough electricity can flow in path 2 because the resistance is much too large. In effect, both pathways don't work. When *heat* is applied to the thermistor its *resistance decreases*. This allows an electric current to flow in path 2 and switch on the thyristor. Path 1 now starts to work, and the buzzer will sound.

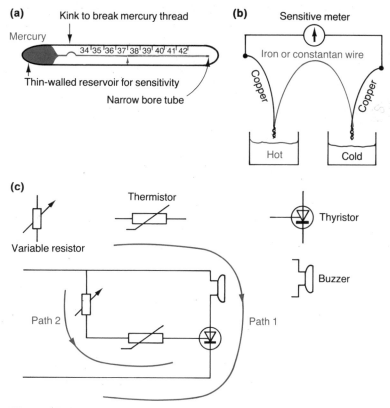

Fig. 16 Different sorts of thermometers

A thermometer's fixed points

The temperatures at which pure water freezes and boils are known as the two fixed points, i.e. 0°C and 100 °C. On an ungraduated (unmarked) thermometer the two fixed points can be marked quite easily.

The lower fixed point
To find the 0°C mark the thermometer should be placed in pure
melting ice. This ensures that the ice is not colder than 0°C.

The upper fixed point
The thermometer should be placed in the steam immediately
above boiling water, which will be at a temperature of 100 °C.
(Impurities raise the boiling point of a liquid.)

Thermostats
The word 'thermostat' simply means 'still temperature'.
Thermostats are devices that are used to control heating circuits
by turning them up/down, on/off. There are several different
thermostats. They basically fall into two types: mechanical and
electrical/electronic.

Mechanical thermostats
The following list of uses for mechanical thermostats is not
complete: gas/electric ovens; central heating control; water
heaters; kettles/irons; car radiators; fire alarms.

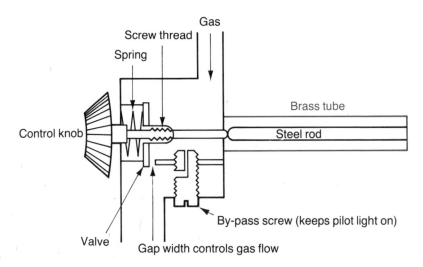

Fig. 17 A gas thermostat – the sort of thing that controls the flow of gas
to a cooker oven.

The gas thermostat (Fig. 17) is controlled by the brass tube contracting and expanding. If the tube cools it pushes the steel rod inwards (to the left) against the spring, pushing the valve further open. The gas flow increases and the oven heats up. This cycle is continuous.

Electric thermostats may use *bimetallic strips*. When one of these is heated it bends. This movement can be used to either switch something on or switch something off (Fig. 18).

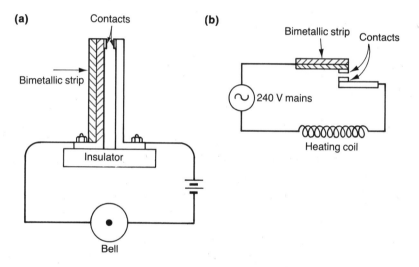

Fig. 18 Bimetallic strips used as: (a) heat-activated thermostat (e.g. fire alarm); and (b) heat-deactivated thermostat (e.g., iron, kettle)

Other practical thermostats may not involve the control of fuels such as gas or electricity. Cars require thermostats to keep the engine at a constant temperature. Most car radiators are fitted with such a thermostat (see Fig. 19). When the engine is cold the valve remains shut and the engine cooling water cannot circulate. When the engine warms up the valve opens and the water flows through the radiator.

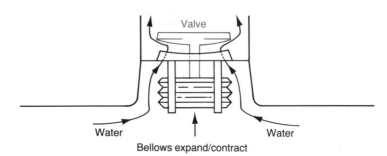

Fig. 19 A thermostat suitable for a car radiator

Electrical/electronic thermostats

These are controlled by electronic circuits or components. A thermistor, as used in the circuit in Fig. 16(c), would work as a thermostat. In that example it causes the buzzer to be activated.

Summary

Energy makes things happen. Energy is measured in joules and is easily changed from one form to another. The final form is always heat. Insulators keep the heat in not the cold out. The energy efficiency of a machine can never be more than 100 per cent. Power (in watts) is the rate of energy transfer.

Practice questions

Try these questions first without reference to the text. Check your answers with the specimen answers at the end of the book.
1 In winter why do air-filled cavity walls keep a house warmer than solid brick walls? (2 sentences). Why does filling the cavity with plastic foam keep the house even warmer? (2 sentences).
2 Explain how a hot water radiator heats up a room.

Aims of the chapter

By the end of this chapter you should be able to:

1 Recall ideas about:
   (a) molecules and the kinetic theory;
   (b) the atomic nucleus and radiations.
2 Distinguish between solids, liquids and gases.
3 Describe experiments to:
   (a) illustrate Brownian motion and diffusion effects;
   (b) show the effects of pressure/temperature on the volume of a gas;
   (c) measure the density of solids and liquids;
   (d) detect radiations.
4 Demonstrate an understanding of radioactivity and nuclear power.

What is matter?

It makes no difference what object or material you think of, it is made of 'matter'. Everything is made of matter. It is what the Universe and everything in it is made of. Sand and cement are two very different kinds of matter. Both these substances appear as small particles, but in very large numbers. Finding out about the differences between the individual particles will tell us more about the different kinds of matter.

Particulate form of matter

It is possible to divide most materials into smaller and smaller pieces. The two examples, sand and cement, are not examples of the smallest kinds of particle. Even a grain of sand (or cement) can be divided. There is a final limit to the number of parts you can cut a piece of matter into without destroying its identity. All matter is made from a vast number of *tiny particles*. This idea is the **particulate nature of matter**.

Particles, molecules and atoms

Copper sulphate is a very common chemical used in school science laboratories. The smallest individual piece or particle of copper sulphate is known as a molecule. The use of the word 'particle' is usually saved for 'a group or (small) number of individual molecules'. A **molecule** is the smallest individual piece of matter that can exist on its own. One single piece of copper sulphate would be known as a molecule of copper sulphate. Suppose you were to take that single molecule of copper sulphate and break it up. It would no longer have its own identity. There are six sections to the copper sulphate jigsaw. These separate sections are called **atoms**. They are the building blocks of all molecules and in turn they are the building blocks of all matter.

Each molecule of copper sulphate ($CuSO_4$) is made from 1 atom of copper, 1 atom of sulphur and 4 atoms of oxygen. The copper sulphate molecule contains six atoms altogether.

Atoms group together to form molecules
A molecule is the smallest individual piece of matter
Molecules group together to form particles

Moving molecules

Robert Brown supplied evidence to support the idea of matter being made of molecules. Fig. 1 shows a typical arrangement for direct observation of the action of small (invisible) particles – molecules.

Holding a lighted straw as shown in Fig. 1 will result in smoke pouring from the lower end. Use this to fill the smoke cell. Leave the smoke cell box connected and ready for operation while you fill the smoke cell. If you use a cover slip, the smoke will stay in the cell for several minutes. Carefully look down through the microscope and adjust the focus. Do not be fooled by: the circular rim of the smoke cell; or the top/bottom surfaces of the cover slip.

As you are looking for **Brownian motion**, you are looking for something moving and not something that is still! When properly adjusted, the view through the microscope will show you some fairly small but bright specks. The specks will be on the move and individual specks will be moving along an uncertain path. This rather 'drunken' movement is known as *random motion*. The bright specks are smoke particles. They are being bombarded by air molecules in the cell.

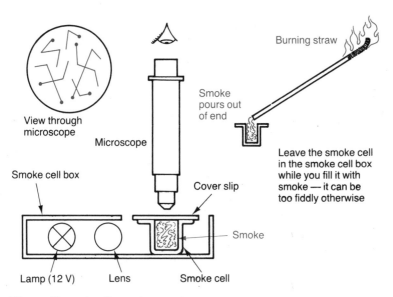

Fig. 1 Observing Brownian motion

Diffusion

Gas molecules are not the only ones to move. Molecules in solids and liquids do, too. A simple experiment to illustrate the movement of these molecules requires only a test tube of water and a copper sulphate crystal. Let the test tube stand for a little while then drop in the crystal. After several hours the familiar blue colour will surround the crystal. Some of the copper sulphate molecules will have *diffused* into the water. The idea of molecules in motion is of the greatest importance. The name given to the idea of moving molecules is the **kinetic theory**. The kinetic theory of molecules helps to explain some other ideas.

States of matter

The phrase *three states of matter* simply refers to whether a material is a *solid, a liquid or a gas*. The three states of matter are: the solid state; the liquid state; and the gas (or gaseous) state. NB Electronics are sometimes referred to as solid state electronics.

A useful material to study the three states of matter is water. Conveniently it exists in all three states – under different names.

State	Common name
Solid	Ice
Liquid	Water
Gas	Steam

(It is possible for the three states of water to exist at the same time. Arctic regions contain a number of icebergs floating in the sea and there is water vapour in the atmosphere above the water.)

Solids, liquids and gases

There is a difference between the three states of the material we know as water. It is all to do with energy (and the amount of energy each state has.) The molecules in ice, water and steam behave in slightly different ways.

The solid (e.g. ice)
The molecules must be arranged in a fixed or regular pattern. They cannot move about by themselves but only in large groups. A typical example would be an ice cube. Whole blocks of molecules stay together and move as one, retaining their shape.

The liquid (e.g. water)
In a liquid such as water the pattern of molecules is not fixed. This is easily shown by turning on a tap and pouring a glassful of water. The water takes up the shape of the container it is in. It *flows easily.*

The gas (e.g. steam)
When a kettle boils the molecules move quickly from the kettle spout and 'fly off' in all directions. They are totally free and can move anywhere. Released into a container, gas molecules will spread out until they occupy the container completely.

The three states and the kinetic theory

Which state a molecule belongs to depends on movement. It depends on how much *kinetic energy (KE)* the molecule has. For a molecule to be a 'solid' it has only a little KE, just enough for it to vibrate about a fixed position. In the liquid state, a molecule has more KE – enough to let it move past other molecules, but not enough to let it break free. Molecules in a gas are very energetic and have large amounts of KE. They have so much energy they can break free of all other molecules and move off on their own.

Expansion

A common mistake to make when thinking about expansion is to say that the molecules get bigger. This is not so!

Expanding solids
When a solid is heated it expands. In other words it takes up more space. Each molecule takes up more space. They take up more space because they have been *given more energy* (more KE). As they have more energy they will vibrate more (but staying in their fixed position) and so need more space. If a solid is given too much energy the molecules will begin to move past one another. The solid will change to a liquid. (Solids only expand by small amounts, but the forces they create can be enormous.)

A simple demonstration of expanding solids is given by the bimetallic strip. Two different metals are joined together to form a two-sided bar. When heated, each side expands by a different amount and the bar appears to bend. When cooled the reverse happens and the bar straightens. If further cooling takes place the bar will continue to bend, but in the opposite direction. Railway lines have joints in them to allow for expansion.

Expanding liquids
The same idea that applies to solids applies to liquids. If *more energy (KE)* is given to the molecules of a liquid they will *move around more and so need more room*. If they are given too much energy they will separate completely. The liquid will change to a gas. If a series of flasks with narrow bore tubes inserted in their bungs are filled with different liquids and placed in a water bath, the different expansion rates of the liquids can easily be seen. As the flasks heat up so the expanding liquids will rise up the tubes. (Liquids only expand by small amounts, but more than solids.) Some thermometers use expanding liquids.

Expanding gases
The story is the same for gases. More energy (KE) means that gas molecules will moves that much faster and so they too will need

more space. If there's nothing else to stop them, the molecules in a gas will move until they completely fill the space they're in (even outer space). Gases can expand by enormous amounts. Of the three states, a gas is really the only one that can be compressed or squashed into a smaller space. An empty flask with a capillary (very narrow) tube in a bung can be used to demonstrate the expansion of gases. Hold the flask upside down, with the capillary tube under water, and gently warm the flask with your hands. The gas inside will expand and bubbles will emerge from the capillary tube.

Changing the volume of a gas

There are two methods for changing the volume of a gas. It can either be *heated/cooled* or *expanded/compressed.*

Absolute temperatures

These are used when describing changes to gases due to heating. The Kelvin (or Absolute) scale has the same size measurements as the Celsius scale, but a different starting point. The following conversions allow for the transfer between scales:

Celsius to Kelvin (or Absolute)   *add* 273
Kelvin (or Absolute) to Celsius   *subtract* 273
e.g. 100 °C → (100 + 273) = 373K
    100K → (100 − 273) = −173 °C

Absolute zero is the temperature at which molecules appear to cease vibrating altogether!

Charles' law

Heating a gas will cause it to expand and cooling it will cause it to contract (shrink). As the volume increases, the Absolute temperature also increases. The connection between volume and Absolute temperature is quite straightforward. If the Absolute temperature of a gas doubles, the volume will double, and so on. The equation linking volume and absolute temperature is:

$$\frac{V}{T} = \text{const}$$

This is Charles' law. The apparatus used to investigate Charles' law is shown in Fig. 2(a). The height of the air column is used to indicate how the volume changes. A series of temperature and height measurements are taken and should be plotted on a graph similar to the one in Fig. 2(b). Care is needed with the heating of the beaker of water. This should be done slowly. The heat source should be removed a short while before any measurements are

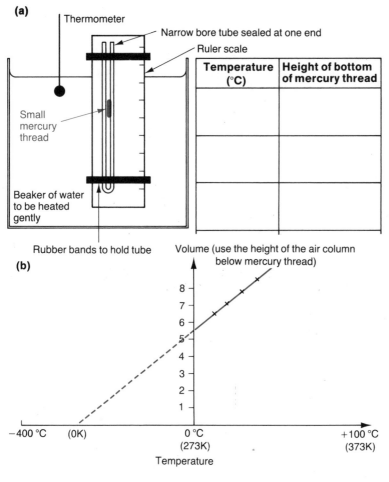

(a)

Thermometer

Narrow bore tube sealed at one end

Ruler scale

Temperature (°C)	Height of bottom of mercury thread

Small mercury thread

Beaker of water to be heated gently

Rubber bands to hold tube

(b)

Volume (use the height of the air column below mercury thread)

8
7
6
5
4
3
2
1

−400 °C   (0K)   0 °C (273K)   +100 °C (373K)

Temperature

Fig. 2 Checking Charles' law

taken. The water should be stirred regularly. The graph is plotted on slightly unusual axes. This is done to allow you to *extend the graph* (the dashed portion) backwards to where it crosses the temperature axis. At this temperature the volume will (in theory) have shrunk to zero, the idea being that if molecules have almost no volume they cannot be moving – they will have no KE. At this point their temperature is as low as it can get, Absolute zero. This experiment allows you to investigate the value of Absolute zero.

Boyle's law

The second option for changing the volume of a gas involves squashing it into a smaller space or letting it expand into a larger one. In other words, increasing the pressure on the gas will cause the volume to decrease. The reverse is true and if the pressure on a gas is reduced, then the gas will expand:

If P increases, V decreases
If V increases, P decreases

These are connected by the equation:

$P \times V = const$

For example, if the pressure doubles the volume shrinks to half, if the pressure halves the volume will double. The apparatus, known as Boyle's law apparatus, is shown in Fig. 3.

  Air from a pump puts pressure on a column of oil and this compresses the air layer at the top of the tube. Pairs of measurements (pressure/volume) can be taken as the pressure increases and the volume decreases. They can then be checked (and confirmed) when the pressure is allowed to decrease again. The *P – V graph is a curve but the P – 1/V graph is a straight line starting at the origin.*

Density

If you had one box of dry sand and an identical box of wet sand, the dry sand would not have the same mass as the wet sand even though they would take up the same space. The idea linking the *same volume* with a *different mass* in kg is called density. Density is used to compare materials to see how their masses (in kg) differ even though their volumes stay the same.

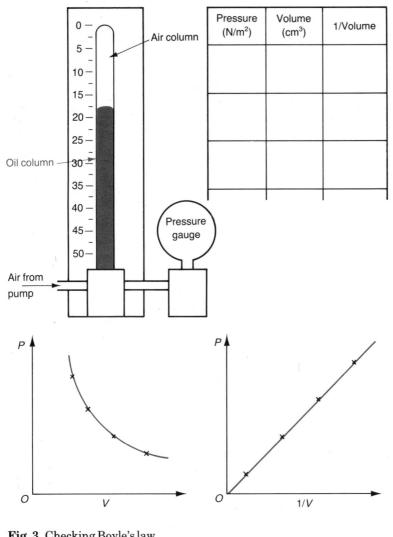

Fig. 3 Checking Boyle's law

Density is the mass of one (unit) volume of a material:

$$\text{Density} = \frac{\text{Mass}}{\text{Volume}}$$

The units of density are mass/volume (kg/m^3). It is sometimes more convenient to talk about grams and cubic centimetres. The units then become g/cm^3.

Steel ships are not solid steel, they are a mixture of steel, air and several other materials. The combined density value of everything on a 'steel' ship is somewhere between $7800\ kg/m^3$ (steel) and $1.3\ kg/m^3$ (air), but it must be less than $1000\ kg/m^3$ (water) if it is to float.

Measuring density

The two most convenient methods for measuring the density of solids are: by direct measurement; and by displacement.

Direct measurement
This is most suitable if the object is regular, the shape of a match box, for instance. To find the volume of the solid measure the three sides and use the equation:

Volume = Length × Breadth × Height

Measure the mass on a set of scales. Use the density equation to find the density value for that particular object/material. Fig. 4(a) shows an example.

Volume = 6 cm × 2 cm × 12 cm = 144 cm^3
Mass   = 100 g

$$\text{Density} = \frac{\text{Mass}}{\text{Volume}} = \frac{100\ g}{144\ cm^3} = 0.699\ g/cm^3$$

Displacement
This very simple idea allows you to find the density measurement of an irregularly shaped object. Pour a convenient amount of water into a *measuring cylinder* (say 100 cm^3). Lower the object under investigation into the water; the water level will rise. If you subtract the original volume from the new volume, you will have the object's volume, in cm^3. If the object is very awkward it may be necessary to fill a *displacement can* (see Fig. 4(b)) with water and then let the overflow fill a measuring cylinder to obtain the volume directly. Mass is measured on a set of scales and the density value found using the density equation.

(a)

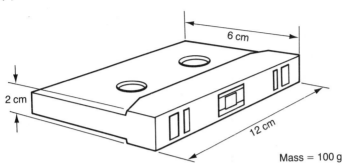

(b)

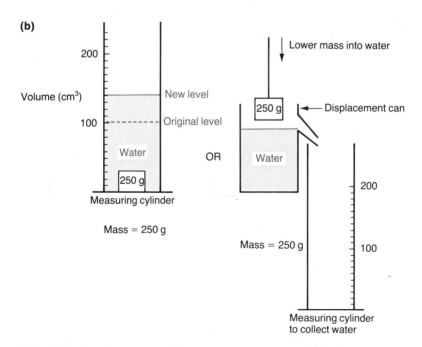

Fig. 4 Finding density by (a) direct measurement and (b) displacement

The example from Fig. 4(b) shows a displaced volume of 40 cm^3.

The mass = 250 g
Volume   = 40 cm^3

$$\text{Density} = \frac{\text{Mass}}{\text{Volume}} = \frac{250\ g}{40\ cm^3} = 6.25\ g/cm^3$$

The atom and its nucleus

There are many different atoms, all having different characteristics. The differences between atoms are due to their internal make-up. An atom is thought to consist of two separate sections: an outer set of shells; and an inner core or nucleus.

Fig. 5 shows how an atom is thought to be constructed. Atoms contain a selection of three things: electrons; protons; and neutrons.   They contain them in different numbers and it is this variation which makes one atom different from another. Table 1 gives the properties of electrons, protons and neutrons.

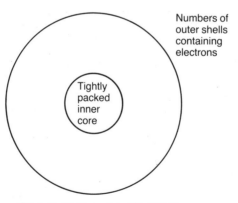

Numbers of outer shells containing electrons

Tightly packed inner core

The outer shells contain **electrons**

The inner core or nucleus contains **protons** and **neutrons**

Fig. 5 Basic atomic structure

	Mass[*]	Charge	Affected by magnetism?	Found in
Electron	1	−1	Yes	Shells orbiting nucleus
Proton	2000	+1	Yes	Tightly bound in nucleus
Neutron	2001	0	No	Tightly bound in nucleus

Table 1 The make-up of atoms   * Approximate values
The shorthand symbols are: proton = $_1^1$p; neutron = $_0^1$n; electron = $_{-1}^0$e

Atomic structure

Using electrons, protons and neutrons it is possible to make up
any kind of atom. Helium (symbol He), for example, needs 4
electrons, 2 protons and 2 neutrons. Carbon (symbol C) would
require 6 electrons, 6 protons and 6 neutrons. An atom always
has the same number of electrons in its shells as it has protons in
its nucleus.

Atomic nuclei

Inside the nucleus the number of protons is simply called the
proton number (Z). The **nucleon number** (A) – often known as
mass number – is equal to the number of protons added to the
number of neutrons (N). This information can be recorded very
simply too. For example, in the nucleus of uranium (symbol U)
there are 92 protons and 146 neutrons.
    For uranium:

The nucleon number = 92 + 146 = 238
The proton number = 92

This is written in shorthand using the chemical symbol U for
uranium:

(nucleon number) $_{92}^{238}$ U (chemical symbol)
(proton number)

The shorthand for the sodium nucleus would be:

$$_{11}^{23}\text{Na}$$

There would be $(23 - 11) = 12$ neutrons in the nucleus along with
11 protons, making a total of 23 nucleons altogether.

Nucleon number = Proton number + Neutron number
(A) (Z) (N)

Chemical reactions and isotopes

It is the number of protons (which is equal to the number of electrons in the surrounding shells) which is responsible for deciding how an atom will react. The neutrons just add bulk and make the nucleus heavier and bigger. In some cases it is possible to take the nucleus of an atom and add an extra neutron or remove an existing neutron. As nothing would be happening to the number of protons, nothing would happen to the way the atom behaves chemically. If the number of neutrons in a nucleus is changed then the new nucleus is called an **isotope**. For example carbon can absorb an extra neutron to become a carbon isotope, and yet still be chemically the same!

$$^{12}_{6}C + ^{1}_{0}n = ^{13}_{6}C$$

Radiation

If a nucleus of an atom is unstable, it will want to break up into smaller more stable portions. Like most things when they break, they don't break cleanly, but give off a number of fragments too. Atomic fragments are called *radiations* because they are emitted, or given off, from the nucleus when it breaks up. These radiations are extremely small, much smaller than the atoms themselves. Because they are so small, they are very difficult to stop.

There are three varieties of radiation emitted from the nucleus: alpha (α) radiation; beta (β) radiation; and gamma (γ) radiation. A fourth radiation, neutron (n) radiation, is also sometimes emitted, but it is not required for GCSE except briefly in the section dealing with nuclear fission and nuclear reactors.

Properties of the three radiations are included in Table 2.

Radiation	Charge	Deflected by magetism?	Typical penetrating power[*]
Alpha (α)	+2	Yes	Few centimetres of air
Beta (β)	−1	Yes	3 mm of aluminium
Gamma (γ)	0	No	Many centimetres of lead

[*] Varies with the strength of the source
Table 2 Properties of radiations

The charges ($+2$) and (-1) are 'opposites'. Alpha and beta radiations are *deflected* in opposite directions by a *magnetic field*.

Background radiation

If a material naturally gives off any of these radiation fragments then it is said to be radioactive. Uranium-238 and radon-226 are two examples of naturally occurring radioactive elements. These and other radiations, especially those from the sun, are continually bombarding all parts of the earth. This natural level of radiation is known as background radiation.

Detection

Because radiations are so small, they are impossible to detect by normal methods such as taste, touch, hearing, sight or smell. There is no way in which human beings can use their own senses to detect these radiations. The most convenient detector is the (wrongly called) Geiger-counter. When a radiation detector (the *Geiger – Müller tube*) is connected to a simple counter, detection of radiation becomes possible (see Fig. 6).

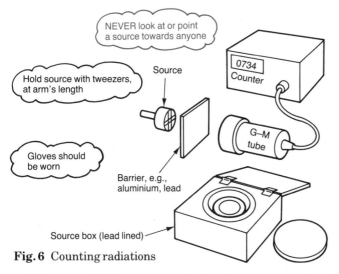

Fig. 6 Counting radiations

A Geiger – Muller tube is a device that emits a pulse of electricity each time radiation passes through it. The counter increases by one with each pulse.

A second form of detector is the cloud chamber (See Fig. 7). When alpha and beta radiations pass through the cooled atmosphere of a cloud chamber, they make white tracks. These are easily seen when illuminated.

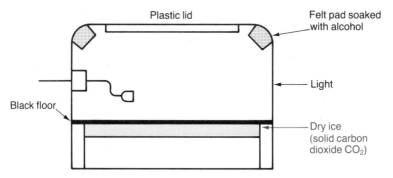

Fig. 7 A cloud chamber

Radiation demonstration

The use of radioactive sources, such as strontium-90, in schools is restricted to demonstration only. Fig. 6 shows a typical arrangement of apparatus suitable for detecting radiations and illustrating their *powers of penetration*. A count of the background radiation should always be made first. As the source is moved closer to the Geiger – Muller (G – M) tube, the count on the ratemeter or counter will increase. Testing the penetration power of the three radiations is done by introducing a barrier between the source and the G – M tube. When the count has reduced to that of the background value then the radiation from the source has been effectively stopped. Alpha particles are stopped quite effectively by a few centimetres of air. Many centimetres of lead are usually required to stop gamma radiation. At the end of the demonstration the background count should be taken again.

The three radiations travel in straight lines from the source. If a source is pointed at a G – M tube they will travel directly. However, if the source contains alpha or beta radiations and a magnetic field is brought near to the radiation path then these radiations will be *deflected* and their directions changed. This effect does not happen with gamma radiation.

Radiations are dangerous because they are very energetic and penetrating. However, once they have lost their energy, they are harmless. A common mistake is to assume that, for example, the barriers used in the radiation demonstration become radioactive. They do not.

Radioactive decay and half-life

Radioactive decay is a *random process*. It is not possible to predict when a particular radioactive atom or group of radioactive atoms will break up and give off their radiations. It is also not possible to make the radioactive decay process speed up or slow down. It is totally random. Some radioactive materials may take years to decay to nothing, whereas others may decay completely in seconds. It is not possible to measure or predict the time it will take for any particular atom or group of atoms to decay. What is possible is to estimate how long it will take for half the atoms in a sample to decay. Fig. 8 is an example of a *decay curve*.

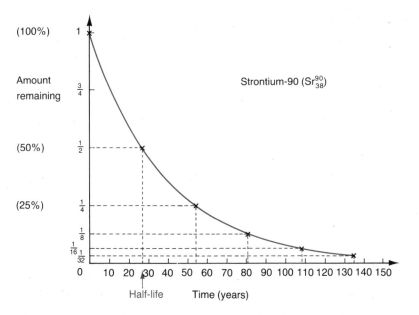

Fig. 8 Decay curve for strontium-90, showing half-life value of 27 years

The time taken for a sample (any size) of strontium-90 to decay to *half its original activity* is 27 years. What remains after 27 years will take a further 27 years to decay to half of this value and so on. The value of 27 years is the **half-life** of strontium-90. Suppose the count rate from the strontium-90 example were 150 counts per minute, it would take 27 years for the count rate to reduce to 75, and so on. Each radioactive material has its own value of half-life. Half-lives vary in length from less than a second to tens of thousands of years.

Uses of radioactive materials

There are several peaceful uses for radioactive materials: the use of radioactive 'tracers' in medicine; using gamma rays to sterilize food to keep it fresh; using gamma rays to sterilize surgical instruments; as a thickness/quality monitor; in oils to test lubricants; and to test the take-up of fertilizers in plants.

Nuclear fission

Nuclear **fission** is the *breaking up* of a large nucleus into a number of smaller nuclei. Some neutrons are also emitted. If these neutrons hit other nuclei then a *chain reaction* begins. Uranium-235 is the example used in Fig. 9. If the chain reaction is uncontrolled, then an explosion may result. A controlled chain reaction forms the basis of nuclear reactors. The chain reaction is

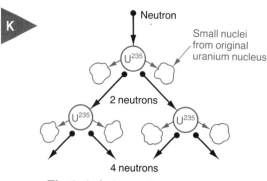

Fig. 9 A chain reaction

controlled by rods known as **moderators**. In the event of an
emergency the control rods fall into the reactor and 'mop up' the
spare neutrons and the chain reaction stops. A chain reaction
creates an enormous amount of heat. This is used to turn water
into steam, which generates electricity in a steam turbine. The
whole reactor is cased in protective concrete. Fig. 10 shows a
typical reactor layout.

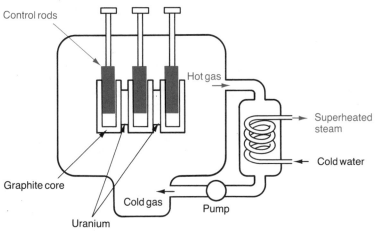

Fig. 10 A nuclear reactor core

Radioactive waste

Because of the dangers of radioactive waste, it has to be stored
with great care. Much waste is kept in sealed units underground
in remote places. Because some of the radioactive materials have
very long half-lives, the waste may have to be stored for hundreds
of years. Some of the 'low level' liquid waste is dumped at sea.

Summary

Matter is made from atoms and molecules. Brownian motion
provides evidence for the particulate nature of matter. The
kinetic theory is used to describe the movement of molecules and

can explain ideas such as expansion and contraction. The nuclei (containing neutrons and protons) of certain atoms undergo radioactive decay by emitting penetrating radiations which are not easily detected.

Practice questions

Try these questions first without reference to the text. Check your answers with the specimen answers at the end of the book. Limit your reply to about the length indicated in the brackets.

1 When a bicycle pump is used to blow up tyres, on the down stroke, what happens to:
(a) the pressure of the gas in the pump? (2 sentences)
(b) the spacing between the molecules? (2 sentences)
(c) the density of the gas? (2 sentences)

2 (a) It is intended to build a nuclear power station near the village of Grimsthwart. Write a notice to the residents of the village as if you were from the proposed nuclear power station (about 10 lines).
(b) Reply to the notice as if you were a resident of the village (about 10 lines).

Aims of the chapter

By the end of this chapter you should be able to:

1 Recall:
   (a) that sound is the vibration of molecules;
   (b) ideas about frequency, wavelength and velocity.
2 Distinguish:
   (a) between transverse and longitudinal waves;
   (b) real and virtual images;
   (c) reflection, refraction and diffraction.
3 Describe experiments to:
   (a) measure the speed of sound in air;
   (b) show reflection, refraction and diffraction;
   (c) disperse white light;
   (d) find the virtual image produced by a plane mirror.
4 Demonstrate a practical knowledge of the application of waves.

What are waves?

It is almost impossible to give a precise description of a wave or set of waves and keep the explanation simple. However, it is very possible to use a model to explain how they work and what they do. Waves are a method of *energy transfer*. Waves carry energy from place to place. They are an energy form. The model used to describe the action and behaviour of a wave is a moving pattern or shape. The repeating pattern model idea is very convenient, except that there are two kinds of repeating patterns. Unfortunately, both repeating patterns are called waves and so it is all too easy to mix up information about them. They are constructed, or made, in totally different ways from one another, but they have certain kinds of behaviour in common. It is important to realize just what the two kinds of wave have in common, and how they are very different from one another.

Transverse and longitudinal waves

The first of the repeating pattern models is the **transverse wave**. Transverse means *across*. A transverse wave moves across the direction in which the wave is travelling (see Fig. 1(a)). Holding on to one end of a slinky coil and jerking it across or sideways (a transverse movement) sends an energy wave along to the other end of the slinky coil. The coil moves in one way, but the energy wave moves in a different direction.

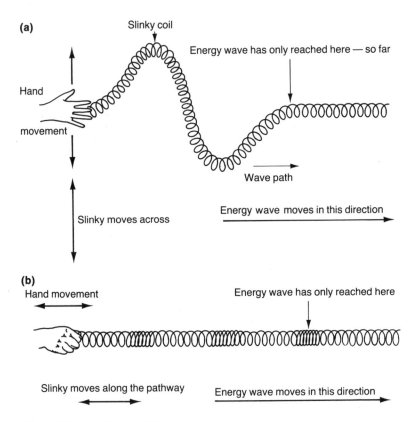

Fig. 1 Wave motions. (a) Transverse – across the pathway. (b) Longitudinal – along the pathway

Any wave motion you can think of belongs to this set of transverse waves with one exception – sound. The most important group of transverse waves is known as the **electromagnetic spectrum** (EMS).

Longitudinal waves only have one set – **sound waves**. A longitudinal wave can also be demonstrated easily with a slinky coil (see Fig. 1(b)). A longitudinal wave is made by moving the slinky coil in a backwards/forwards motion at the held end. This produces a compression in the coil. The energy is carried by this compression as it travels to the end of the slinky.

K

The compressions in a longitudinal wave are created by molecules being bunched together. They pass on energy in the form of vibrations.

The properties of different types of waves are summarized in Table 1.

Waves	Longitudinal waves (only) sound	Transverse waves All EMS waves	Water (surface)
Move across the pathway	×	√	√
Can travel through a vacuum	×	√	–
Move along the pathway	√	×	×
Need a material to travel in	√	×	–
Carry energy	√	√	√
Can be reflected	√	√	√
Can be refracted	√	√	√
Can be diffracted	√	√	√

Table 1 Different types of waves share some things in common, but not others

Wavelength

Fig. 2 shows two sets of waves. The distance between repeating parts of the pattern (the length of each single wave) is known as the **wavelength** (λ). Wavelengths are measured in metres (m). A wavelength can be measured from any point on the wave to the next identical part of the wave. The wave in the left of Fig. 2 has a longer wavelength (3 m) than the wave alongside it (2 m).

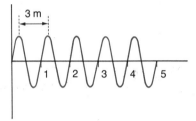

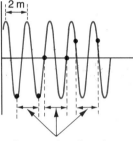

One wavelength each

Fig. 2 Different wavelengths

Frequency

Frequency is a measure of *how often* something happens. The symbol for frequency is *f* and the units are **hertz** (Hz). Something that occurs 10 times every second has a frequency of 10 hertz (Hz).

Travelling waves

The speed of a set of waves can be found in two ways: from information about distance travelled and time taken; or from information about frequency and wavelength.

Distance and time
The relationship between distance, speed and time is exactly the same as that given in Chapter 1, Forces. The equation used to calculate speed is:

$$\text{Speed (m/s)} = \frac{\text{Distance (m)}}{\text{Time (s)}}$$

At school this method is only practical for the measurement of the speed of sound.

Example An echo takes 1.5 seconds to reflect from the front wall of a school building 250 m away. What is the speed of sound in air?

Distance for echo = 2 × 250 m = 500 m; Time taken = 1.5 s

$$\text{Speed} = \frac{\text{Distance}}{\text{Time taken}} = \frac{500}{1.5} = 333 \text{ m/s}$$

Frequency and wavelength

The more waves that pass a point in one second the faster the wave is travelling. The longer the wavelength the faster the wave is travelling.

$$\underset{\text{(m/s)}}{\text{Speed}} = \underset{\text{(m)}}{\text{Wavelength}} \times \underset{\text{(Hz)}}{\text{Frequency}}$$

It is not really practical to use this method for measuring waves belonging to the EMS. (It is more easily done with sound waves.) Neither method is really necessary for GCSE.

Example Capital Radio broadcasts throughout the London area using radio waves with a frequency of 1548 kHz, travelling at 300 000 km/s. To what wavelength (in metres) would you tune a radio to receive Capital programmes?

Speed = 300 000 km/s = 300 000 000 m/s;
Frequency = 1548 kHz (1 548 000 Hz)

The correct form of the equation to use is:

$$\text{Wavelength} = \frac{\text{Speed}}{\text{Frequency}}$$

$$= \frac{300\,000\,000}{1\,548\,000} = 193.8\,\text{m}\,(194\,\text{m})$$

Sound waves

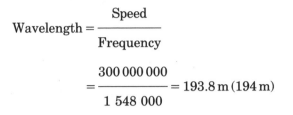

Sound waves are longitudinal waves. They would be more correctly called *sound vibrations*. Sound is the transfer of energy in the form of vibrations. The vibrations are passed on by the molecules in a material. If there are no molecules present, then sound energy cannot exist or be transferred. Space (a vacuum) is silent. A simple demonstration to show this uses a bell inside a jar. When the jar is connected to a vacuum pump and the air removed, sound can no longer be heard. (However, light waves, part of the EMS, can travel through the jar.)

Sound travels much better through some materials than it does others. The more 'solid' a material is the easier it is for molecules to pass on vibrations. This is also true the closer molecules are to one another. The speed of sound in a material depends on: the 'state' of the material (solid/liquid/gas); and the density of the material. Note that this is only a guide, not a rule – there are exceptions.

For example, the speed of sound in steel is over 5000 m/s but in air it is only 340 m/s. Materials such as expanded polystyrene would make good sound insulators because of the way in which they are made, i.e., low density and quite spongy. Sound is medium (material) dependent.

Displaying sound

It is not convenient or practical to 'see' sound as a set of compressions. A more convenient way is to transfer the sound to a *cathode ray oscilloscope* (CRO). This can be done with a microphone. Sound is converted to a series of electrical impulses and these impulses are displayed on the screen. Fig. 3 shows such a conversion. *Do not* mistake the electrical trace on a CRO for a picture of a sound wave. The pattern in each of the three models can be seen to repeat itself.

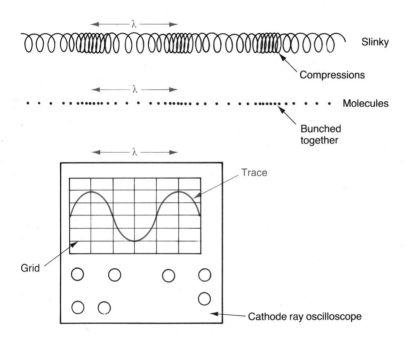

Fig. 3 Three models of sound

Hearing

Human hearing in a young person is sensitive to a range of about 20 Hz to 20 000 Hz. There are sounds above 20 000 Hz; it's just that human hearing cannot cope with them. Many animals can hear these high frequencies. For example, bats and dolphins can hear frequencies up to 100 000 Hz. Frequencies above the human hearing range are known as *ultrasonics*.

Pitch

Plucking different strings on a guitar or pressing different keys on a piano will give different notes. It would be very difficult to tell exactly what frequency they were. However, it would be a fairly simple task to tell which note was the higher frequency. The word 'pitch' is used to describe a higher or a lower frequency. A high-pitched note is further up the frequency scale than a note with a low pitch.

Loudness

In addition to the pitch of a note, loudness is important. It would be quite possible to hear two notes of exactly the same frequency yet be able to tell that one is louder than the other. The difference in strength or loudness of a note can be shown on a wave diagram (see Fig. 4).

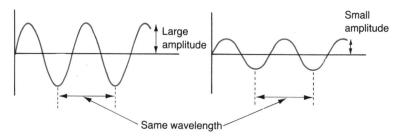

Fig. 4 Same wavelength and frequency, but different amplitude

In Fig. 4, the left-hand wave has the same frequency and wavelength as the right-hand wave but a larger **amplitude** (i.e. the height of the wave). Loudness *depends on amplitude* so the left-hand wave would sound louder. As well as being louder, it would transfer more energy.

Echoes

When a sound wave hits a surface it will be reflected. Surfaces that are more 'solid' make better reflectors. Several practical situations make use of echoes. *Sonar* is a device for displaying information received as sound waves. There are two types of sonar: *active* and *passive*. Active sonar emits a series of sound waves and listens for the echoes reflected. There are applications for active sonar on land, sea and in the air (see Fig. 5).

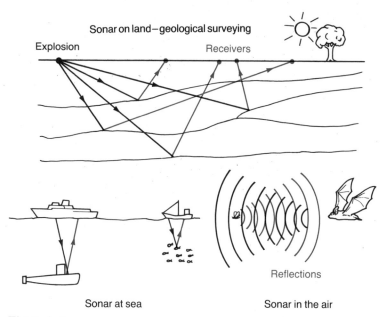

Sonar on land – geological surveying

Explosion   Receivers

Reflections

Sonar at sea   Sonar in the air

Fig. 5 Active sonar

Passive sonar differs from active sonar in one way only. Unlike active sonar it does not emit any pulses, it just listens.

Velocity of sound

There are several methods available for measuring the velocity of sound. The method described here is not the only one, but it is one of the easiest. It requires two people to work together in a team and involves timing an echo.

The side of a building is ideal for this experiment. It should preferably be flat and there should be no other large buildings nearby. The team should walk a convenient distance away from the building wall, say 100 m. Facing the wall directly, one member of the team begins to clap two pieces of wood together. The echo should be heard distinctly. The aim is to clap the wood together in time with the echo. The second team member counts the clapping. At a convenient time to both, the person counting restarts the count, beginning at zero and at the same time starts a stopwatch. The count of clapping and stopwatch should continue until the clapping count reaches, say 50, when the stopwatch is stopped.

Typical results for such an experiment could be:

Distance from wall = 100 m
Time for 50 echoes = 30 s

The velocity of sound is found in the following way:

Each echo = 200 m (to the wall and back)
50 echoes = 50 × 200 m = 10 000 m

Sound has travelled 10 000 m in 30 s.

$$\text{Speed} = \frac{\text{Distance}}{\text{Time taken}}$$

$$\text{Speed} = \frac{10\ 000\ \text{m}}{30\text{s}} = 333.3\ \text{m/s}$$

(The accepted value for the velocity of sound in still air is about 330 m/s.)

This experiment is a simple version of active sonar.

Aircraft that fly in excess of the local speed of sound are said to fly at *supersonic* speeds. They will leave their sound behind them as a *shock wave*.

The electromagnetic spectrum, EMS (light waves)

K

Light waves form only a small part of the EMS. The complete EMS is shown in Fig. 6.

The wavy lines indicate where one variety of waves changes to another type. These are not fixed borders, but are very flexible dividing lines. Their purpose is to divide groups of the EMS that share approximately the same properties. You should be able to place them correctly in order of wavelength/frequency and

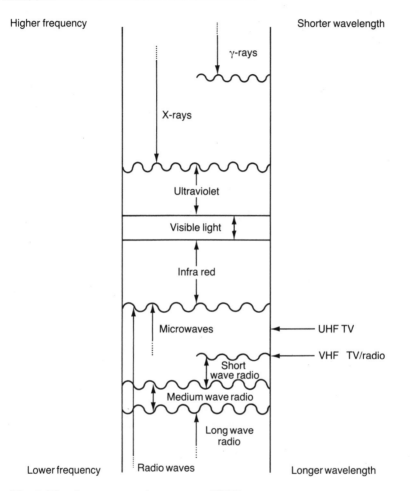

Fig. 6 The electromagnetic spectrum (EMS)

provide practical examples for the use of each major group of
waves. For example: γ rays – cancer treatment; X-rays –
industrial/medical photography; ultraviolet – sun tanning/
photography; visible light – human sight/photography; infrared –
heating effect/burglar alarms/photography; microwaves –
cooking; and radio waves – communication.

NB Microwaves do not strictly speaking cause heating. The best receivers of microwaves are water molecules. Microwaves make the water molecules vibrate more (i.e. they heat up), which makes the food cook.

Travelling waves

Radio waves are a practical example of energy waves that penetrate seemingly solid objects. They can also travel in the vacuum of space. Members of the EMS do not need a material to travel in. They are not medium-dependent like sound waves. All EMS waves travel at the same speed in a vacuum: 300 000 km/s.

Dispersion of white light

The part of Fig. 6 labelled 'visible light' contains several different varieties of light. It is broken up into what are called colours. The *visible spectrum* has seven colours: red, orange, yellow, green, blue, indigo and violet. In fact there are several different greens, several different blues, etc. Each different colour has a different wavelength. The red end of the visible spectrum has longer wavelengths than the blue end.

White light does not exist as a separate 'colour' at all. It is a mixture of all the individual colours. This can be demonstrated by using white light and splitting it up. The splitting up of white light is known as **dispersion**. Fig. 7(a) shows apparatus suitable for dispersing white light.

The red light is bent less than the blue light. The other colours (wavelengths) are in order, between them (see Fig. 7(b)).

It is possible to produce a spectrum without either of the two lenses shown in Fig. 7. This usually results in a smaller less well spread out spectrum.

Light passages and images

Very often phrases like 'light rays', 'light beams' or 'light paths' are used to describe a passage of light to or from an optical device. They are different ways of saying the same thing, but for convenience the following meanings will be applied:

Light ray   Thin passage of light
Light beam   Wide passage of light

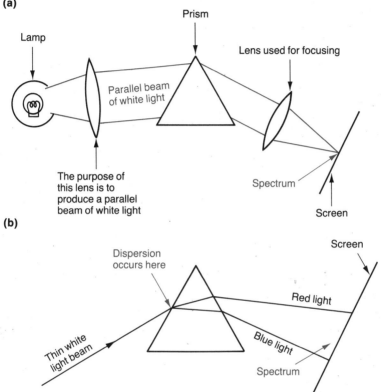

Fig. 7 (a) Producing a pure spectrum from white light. (b) Blue light is bent more than red light

The images produced by optical devices fall into two groups: *real* and *virtual*. Crudely: a real image exists, but a virtual one does not, although the human eye is fooled into thinking that it does.

A real image can be projected onto a screen
A virtual image cannot be projected onto a screen

For a single optical device (1 mirror, 1 lens, etc.) the image produced is always:

Real: if it is upside down compared with the object (e.g. camera lens)

Virtual: if it is the right way up compared with the object (e.g. magnifying glass)

Reflection and images (from a plane mirror)

There are two experiments that are useful in showing the properties of a plane mirror. The first involves measuring the angles at which light strikes a mirror and then rebounds. The second involves the image produced by a plane mirror.

The law of reflection states that '**the angle of incidence** of a light ray is equal to **the angle of reflection** of a light ray'. Checking this idea is very straightforward. A light ray is directed towards a mirror and the reflected ray observed.

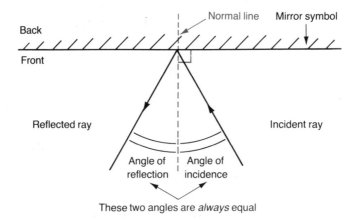

Fig. 8 Angle of incidence equals angle of reflection

In Fig. 8 a dotted line has been drawn in and labelled 'Normal'. This line does not really exist; it is just drawn there to help with the measurement of angles. Any measurement is **always** made from this drawn line. For example, the angle of incidence is the angle measured from the **normal** to the incident ray.

To check the law of reflection an angle value for the reflected ray should be taken at the same time as the angle value for the incident ray. This should be repeated a number of times, e.g.:

Angle of incidence	Angle of reflection
35	35
71	71

The image produced by a plane mirror is always virtual. It is the same way up as the object (even though it is reversed or back to front – known as **lateral inversion**). Locating a virtual image is not possible. Finding where you think it is, is possible. Fig. 9 shows an arrangement suitable for looking at a virtual image produced by a plane mirror.

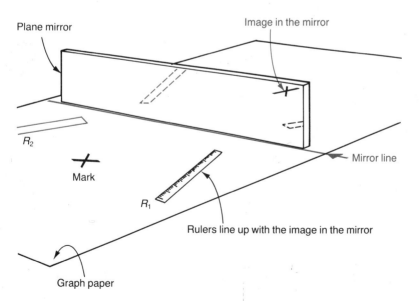

Fig. 9 Finding a virtual image formed by a plane mirror

Support a plane mirror in the centre of a piece of graph paper. Draw a line on the paper at the front of the mirror – this is the mirror line. Draw a mark (X) in front of the mirror. Put a ruler (R_1) flat on the paper. Move it until the image of the mark (X) is directly in line with one edge of the ruler. Draw along this side of the ruler with a pencil. Repeat the ruler exercise a second time (R_2). Both those lines of sight will be directed towards the virtual image seen in the mirror. Remove the mirror from the paper and continue drawing along the two lines until they meet. The position at which the two lines cross is where the virtual image of the mark (X) was formed by the mirror.

The distance from the mark (X) to the mirror line should be the same as the distance from the virtual image position to the mirror line. *Plane mirrors act like lines of symmetry.*

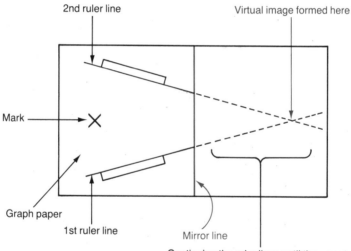

Fig. 10 Locating a virtual image formed by a plane mirror

Refraction

K

When a light ray enters a new material, such as glass, its direction may be changed. **Refraction** is the *changing of direction* of a light ray when it crosses a boundary into another material.

When a light ray, travelling in air, hits a new material straight on (at 90°) there is no change in direction. If it strikes the material surface at an angle, refraction will occur. These two ideas are shown in Fig. 11(a). If there are two boundaries then refraction will occur twice. Sending a light ray at an angle into a glass block will demonstrate this easily (see Fig. 11(b)). No arrows have been drawn on this diagram to indicate direction. This is because *light is reversible* and can go in either direction. (The direction arrows in Fig. 11(a) could also be reversed.) The angles of refraction are clearly labelled (*r*).

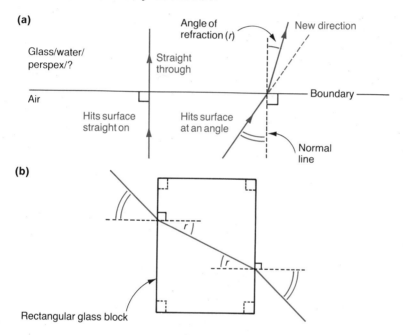

Fig. 11 (a) Refraction happens at a boundary. (b) Two boundaries, two normals, two refractions

Internal reflections

A strange effect happens on the inside of a material such as glass, when light tries to leave it by striking the boundary at a large angle. If this 'striking' angle is too large, the light will not escape from the material but it will be reflected *internally*. The internal angle at which this first occurs is known as the **critical angle**.

The inside surface then acts like a plane mirror (see Fig. 12). It is more convenient to use semicircular glass blocks to show internal reflection.

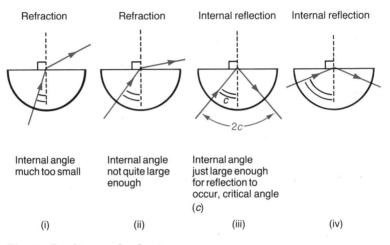

Fig. 12   Total internal reflection

When you do this experiment draw an outline of the block on paper, marking the centre of the flat side. The light ray should always be directed at this point. By increasing the angle at which the light is pointed towards the flat surface the refracted angle will increase. When internal reflection (just) occurs, stop, mark the passage of the light rays entering and leaving the block. Draw lines between these two points and the mark originally put on the paper at the centre of the flat surface. The angle between these two lines is twice the critical angle (because the law of reflection of light applies).

Internal reflection occurs naturally in raindrops and has useful application including optical fibres, periscopes, binoculars, cameras and cats' eyes reflectors in roads.

Curved lenses and mirrors

Lenses and mirrors fall into two groups: **concave** or **convex**. The names refer to the shape of the lens or mirror, not to what they do. The easy way to remember which is which is from the name con*cave* – it *caves* inwards. Fig. 13 shows the various lens and mirror symbols/alternative symbols.

Curved lenses and mirrors have more in common than just shape. A **converging** lens or mirror can bring light together, to focus it, to converge light to a point. **Diverging** lenses and mirrors can only spread light out.

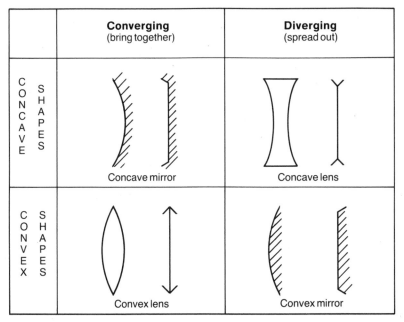

Fig. 13 Lens and mirror symbols

Converge/diverge

Converging lenses and mirrors can *focus a beam* of light at a point. The point onto which a parallel beam of light converges is known as the **focal point**, sometimes called the *principal focus*. If light is emitted from a source placed at the focal point of a converging lens or mirror, then light will emerge as a parallel beam. Examples of this idea include torches, car headlamps, electric bar fires and radar dishes. The distance between the focal point and the lens or mirror is known as the **focal length**.

Lenses let light pass through them when forming images. Mirrors reflect the light to form images (Fig. 14).

Diverging lenses and mirrors cannot form real images, because they are unable to bring the light together necessary to make a real image.

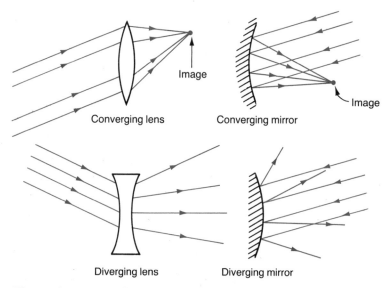

Converging lens   Converging mirror

Diverging lens   Diverging mirror

Fig. 14 Converging/diverging beams of light

The single lens camera

The convex lens of many cameras may be adjusted for focus. This produces a crisp, sharp image on the film when the shutter is opened. The image is real (it is upside down) and smaller than the object (diminished in size). A camera lens produces a *diminished real image*.

The eye

A convex eye lens can be squashed to make it thick, or relaxed to make it thinner. This controls the eye's focus. A real image (upside down) is produced on the retina. It is smaller than the object (diminished in size). An eye lens produces a *diminished real image*.

The projector

A piece of film is used as an object. Light passes through the film, through a convex projector lens and then to a screen. The lens can be moved to focus the image on the screen. The film is inserted upside down to produce an upright image. As the image has been turned upside down (to turn it back up the right way) it must be a real image. It is larger than the object (magnified). A projector lens produces a *magnified real image.*

The magnifying glass

The image seen through a magnifying glass is the same way up as the object. It is a virtual image. A magnifying glass produces a *magnified virtual image.*

Ripple tanks

A ripple tank is a very useful model for studying the properties of waves. The suggestion is simple: if a water wave behaves in a particular way, then sound waves and electromagnetic waves might also behave in the same way. This can also be stated in reverse, because if light and water waves behave in the same way, then this is evidence that light is a wave form. (The same ideas go for sound too.) The general layout of a ripple tank is shown in Fig. 15. Any movement of the water's surface will be projected downwards as a shadow pattern.

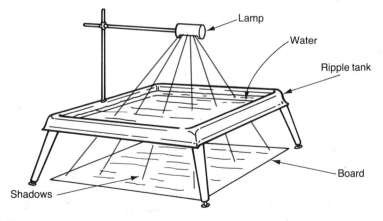

Fig. 15 Ripple tanks provide shadow patterns

Reflection

Fig. 16 shows the reflection of waves from a barrier. If plane waves meet a straight barrier, plane waves are reflected. This is identical with a ray of light being reflected from a plane mirror (angle of incidence equals angle of reflection). The plane waves hitting a curved barrier are reflected as curved converging/diverging waves.

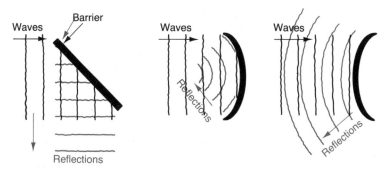

Fig. 16 Reflections at a barrier

Refraction

When a piece of perspex or glass is added to the tank, the water above it will be shallow. Sending water waves through a shallow area is equal to sending light waves through a glass block (see Fig. 17).

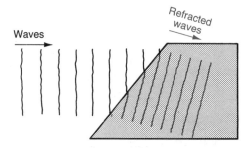

Fig. 17 Water waves are refracted, too

Diffraction

Diffraction is the spreading out of waves when the waves have passed a sharp corner or edge (see Fig. 18). It is not noticeable under normal circumstances with members of the EMS though it does happen. Sound waves can be diffracted quite easily. Most people will have experienced hearing things going on 'round the corner' even though they cannot see what is happening.

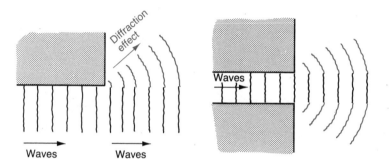

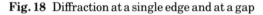

Fig. 18 Diffraction at a single edge and at a gap

Summary

The different types of wave share only certain things in common. All waves transfer energy. The speed of longitudinal (sound) waves is medium-dependent but the speed of transverse (EMS) waves is not. White light can be dispersed into the colours of the spectrum. There are two kinds of image formed by light; real and virtual. Ripple tanks can be used to show reflection, refraction and diffraction.

Practice questions

Try these questions first without reference to the text. Check your answers with the specimen answers at the end of the book.

Answers to questions involving calculations should show all the working and should contain a small amount of explanation including any equations used.

1 Fig. 19 shows two rays of light leaving an object at P and striking the mirror. Draw the two reflected rays and find the virtual image behind the mirror.

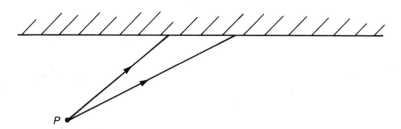

Fig. 19 Where is the virtual image?

2 (a) A fishing boat sends out sonar waves to locate a shoal of fish. The pulse returns after 0.1 s. The speed of sound in water is 14 400 m/s. How far down are the fish?

(b) In practice why will the return pulse last for longer than the pulse originally sent out? (2/3 sentences).

Aims of the chapter

By the end of this chapter you should be able to:

1 Recall ideas about:
(a) static and current electricity;
(b) current, potential difference and resistance.

2 Distinguish between:
(a) static and current electricity;
(b) conductors and insulators;
(c) current, potential difference and resistance;
(d) series and parallel circuits.

3 Describe experiments to:
(a) show the heating effect of a current;
(b) measure resistance using ammeters and voltmeters correctly;
(c) show the effect of width and length on the resistance of a material.

4 Demonstrate:
(a) a practical knowledge of the supply and distribution of electricity;
(b) a knowledge of electrical hazards.

Introduction

You will probably be familiar with the phrases 'static electricity' and 'current electricity'. From this you might imagine that there are two kinds of electricity. It is worth noting now, at the beginning of this chapter, that there is no difference in the electricities at all. They may appear to be different (on the surface) but they are not. The word 'static' means 'still'. Electricity which is 'static' is still and does not move. Make exactly the same electricity move and the 'static' becomes 'current'. Current means 'flowing'.

Insulators and conductors

Materials usually fall into one of two groups: conductors or insulators. (There is a small, but very important, third group called semi-conductors, and they will be dealt with in Chapter 6.) A **conductor** of electricity is a material that allows electricity to flow through it. An **insulator** will not allow electricity to flow through it. At the same time, because an insulator will not allow the passage of electricity it may act as a store for electricity. Electricity which is *static* will stay on an insulator because it has no way of escaping or leaking away. Examples are:

Insulators	Conductors
Wood	Metals
Plastic	Human beings
Ceramics	Tap water*

Semi-conductors (see Chapter 6)

* Pure water is a good insulator but as there are always impurities in water it is safer to think of it as a conductor.

Static electricity

Friction is usually the cause of static. Rubbing a balloon on a woollen pullover will allow you to 'stick' the balloon to a wall. When an object has some static electricity on it then it is said to be charged or to have a static charge on it. There are two types of static charge; *positive* (+ve) and *negative* (−ve). An uncharged piece of material has an equal number of each, so there appears to be no charge at all. Friction between materials causes a transfer of charge from one material to another. When the two are separated, one material will be positively charged and the other will be negatively charged. It is only the negative charges that move. A good example of this would be a cloth duster and a polythene rod. When they are rubbed together and then separated, both end up being charged. The polythene is an insulator and so the charges will stay. This can easily be demonstrated by using the charged rod to pick up small pieces of paper. Some simple experiments involving static charge (created by friction) are shown in Fig. 1.

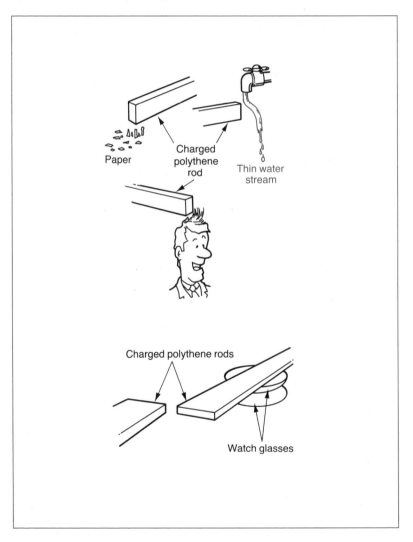

Fig. 1 Experiments showing attraction or repulsion caused by static electricity

Attraction and repulsion

Static charges share a lot in common with magnetism and will: attract other materials which are uncharged; attract other materials which have the opposite charge; and repel other materials with the same charge.

The only true test for an object being charged with static is *repulsion*. Then you are sure that both objects (including the one being investigated) are charged.

The lower part of Fig. 1 shows a suitable arrangement for checking repulsion. When you try this experiment it is important that you do not touch the charged end of either rod or the experiment may fail.

Examples of static electricity are the charges built up on: television screens; carpets (especially nylon); clothes (especially nylon); clouds (lightning); petrol tankers; and dust on records.

Static can usefully be employed in: chimneys (to attract the smoke particles); electrostatic photocopiers; and painting of car bodies.

Current electricity

Whenever static moves then it is called *current electricity*. When you touch an object that is charged, you will provide a pathway for that electricity. An electric current (symbol *I*) will flow through you to earth and you will receive an electric shock. Providing this alternative pathway is known as *earthing*.

More normal sources of current electricity include batteries, dynamos, etc.

An electric current needs a *complete circuit* to flow in. If there is a break in a circuit (anywhere) the current flow will cease.

Circuits and circuit symbols

In studying electricity/electronics as a topic it will be necessary to make or draw electric circuits. In practice, wire tends to get everywhere and people panic. Practical circuits often look nothing like their circuit diagrams. A simple example of this is shown in Fig. 2. Often it is easiest to build a circuit by starting at one side of the source (battery or power-pack) and adding the

various parts one after another – in order, finishing up at the
other side of the source. Using your finger as a pointer is often a
useful method of checking a circuit. Follow the circuit diagram
around then do the same with the practical circuit.

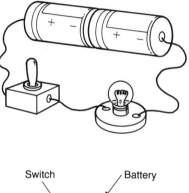

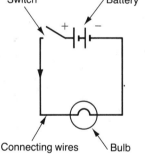

Fig. 2 A practical circuit may look nothing like the circuit diagram

There are several circuit symbols to be familiar with. The ones
already shown in Fig. 2 need no explanation because they have
been drawn in a practical situation too. You need to be able to
translate between practical circuits and circuit diagrams. A full
list of circuit symbols is given in Appendix 1 at the end of the
book.

Sources

Direct current (dc) sources supply electricity in one direction only. All dc sources should have their terminals clearly labelled positive (+ve)/negative (−ve). It is accepted that dc currents flow in the direction positive to negative. This is known as conventional current. An arrow indicates the direction of current flow when the circuit is switched on.

Current is measured in amperes

It is possible to measure 'how much' electricity is flowing in a wire using an **ammeter**. The unit of current flow is the **ampere**, often shortened to 'amps' or 'A', e.g. 5 amps, 2 A etc.

Current is not lost

K A mistake, which is made all too often, is to assume that current is lost as it moves around a circuit. Whatever current flows into a

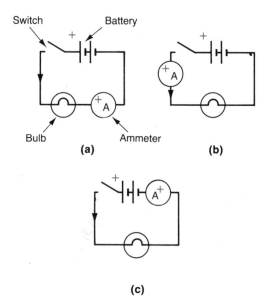

Fig. 3 The ammeter readings will be the same in each circuit

device must flow out again. Similarly, any current leaving one terminal of a source must return to the other (see Fig. 3).

Using ammeters

Like many electrical devices, ammeters suitable for use with dc will usually only work one way round. If you try to run one in reverse, you may damage it. The circuits shown in Fig. 3(a), (b), (c) are electrically identical and any of them would be suitable for measuring the current flowing through the single bulb. (The ammeter reading should be the same in each case. This experiment is suitable for demonstrating that current is not lost around a circuit.) The positive (usually red) connection of the ammeter is always connected to the positive side of the supply.

Kirchhoff's law

At any junction (of wires) the total current value flowing into a junction must equal the total current flowing away from the junction: for example, if a total of 6 A flows into a junction then 6 A must flow out of the junction (see Fig. 4). (It doesn't matter how many wires are connected to the junction.)

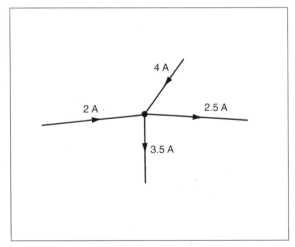

Fig. 4 Kirchhoff's law at a junction

Series and parallel circuits

The circuit diagrams in Figs. 2 and 3 are **series** circuits. All the components (wires, bulbs, etc.) follow on one after another in a single loop or chain. There is a second type of circuit known as a **parallel** circuit. These have components laid out and connected side by side rather like parallel railway lines. It is very unusual to have a completely parallel circuit. More often a circuit is a mixture of parallel and series sections.

In Fig. 5 the parallel section (containing 2 bulbs) is really in series with some connecting wires, an ammeter, a switch and a battery. It would still be called a parallel circuit.

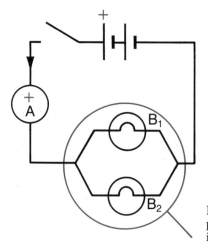

Fig. 5 Circuit with a parallel section (ringed in colour)

The parallel circuit works because of Kirchhoff's law (see Figs. 4, 6). Some of the electric current flows into bulb B_1 and the remainder flows into bulb B_2. These currents meets up again on the far side of the parallel section. Do not assume that an equal amount of current flows through each bulb. This is only the case when bulbs are identical.

One advantage of parallel circuits is that if one part of the parallel section fails to work it has no effect on the other. In Fig. 5, one of the bulbs could go out without affecting the other. If the two bulbs had been in series and one had failed the other would also go out, because there would no longer be a complete circuit.

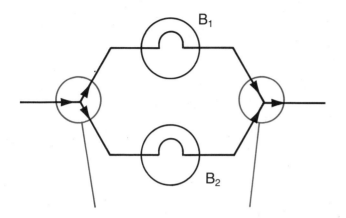

Fig. 6 A parallel circuit. Kirchhoff's law operates at the two junctions ringed in colour

Effects of electrical current

K

There are three main effects caused by an electric current flowing through a device: magnetic effect; heating effect; and chemical effect.

Magnetic effect
Whenever electricity flows through a conductor, a magnetic field is created around that conductor. This is known as electro-magnetism (see Chapter 7).

Heating effect
The passage of electricity through a material has a heating effect on the material. Some materials heat up more than others. For this reason fuses are included in many circuits. They are designed to heat up and melt (or break) before the other materials in the circuit overheat.

Chemical effect
When electricity passes through some materials, it causes a chemical reaction to occur. Electrolysis/electroplating are examples of this effect. The reverse is also true; a mixture of certain chemicals will produce an electric current when they react together. A battery uses this effect.

Resistance

The term **resistance** is used to describe any object/device/
material that *opposes the flow* of electricity. Many questions or
examples will refer to resistance in a circuit. In practice, these
resistances could quite easily be practical devices such as TVs,
washing machines, etc. It's just more convenient to call them a
simple resistance and label them R, for example.

A large resistance will allow less current to flow than a small
resistance. Resistance is measured in **ohms**, symbol Ω.

Resistance wire

Copper has a low resistance and so it is a very good conductor.
Other materials such as constantan or nichrome have high
resistances. These last two types of wire are suitable for
investigating the effect of length/thickness of a material on
resistance. Fig. 7 shows a suitable arrangement for this purpose.

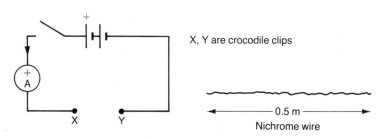

Fig. 7 Resistance changes with length and thickness

Connect the 0.5 m length of nichrome wire between X and Y
and record the ammeter reading. Move one of the crocodile clips
(X or Y) to half way (0.25 m) along the wire and take a second
reading from the ammeter. Disconnect the crocodile clips and fold
the wire in half. Reconnect the crocodile clips (X, Y) and take a
third ammeter reading. The following trend should be quite
noticeable:

Reading No.	Wire length	Single/double	Ammeter reading
1	0.5 m	Single	Low
2	0.25 m	Single	Medium
3	0.25 m	Double	High

Resistance increases with increasing length.
Resistance decreases with increasing thickness.

Resistors

These are devices which have no other function than to 'resist' the flow of current. Individual resistors are of two types: fixed or variable. Fixed resistors are colour coded and are available in a number of resistance values. Variable resistors (*rheostats*) do exactly what they suggest; they allow you to vary the resistance and so control the amount of current flowing through them. A volume control on a radio is one example of the use of a variable resistor. Resistors allow you to limit/control the current flowing in a circuit.

Resistance in series

When resistances are laid out in series in a circuit the total resistance increases. Resistances add together in the following way:

$R_{total} = R_1 + R_2 \ldots \ldots$
R_{total} is the total resistance of the circuit
R_1, R_2, etc., are the individual resistances.

Example   What is the total resistance of the following three resistors in series: 30 Ω, 40 Ω, 1000 Ω?

$R_{total} = R_1 + R_2 + R_3$
$= 30 + 40 + 1000 = 1070\,Ω$
Total circuit resistance = 1070 Ω

Resistances in parallel

A set of resistances in parallel will allow more current to flow through them than if the same resistances had been arranged in series. The total resistance seems to decrease (see 'Resistance wire', on page 108).

For two resistances in parallel their total resistance is found from the following equation:

$$R_{total} = \frac{R_1 \times R_2}{R_1 + R_2}$$

Example   What is the total resistance to two 25 Ω light bulbs arranged in parallel?

$$R_{total} = \frac{R_1 \times R_2}{R_1 + R_2}$$

$$= \frac{625}{25 + 25} = \frac{625}{50} = 12.5\,\Omega$$

Total circuit resistance = 12.5 Ω. (This is considerably less than either of the individual bulbs. Had they been connected in series the total resistance would have been 25 + 25 = 50 Ω.)

Potential difference (pd)

An electric current needs a *driving force* to make it move from place to place. The driving force which makes an electric current flow around a circuit is known as **potential difference** (pd). A pd is always measured between two points in a circuit. Potential differences are measured on a voltmeter. The units are **volts** (symbol *V*). Fig. 8 shows how to connect a voltmeter into a circuit. This is always done last. Do not put the voltmeter in the circuit until the remainder is complete. Voltmeters measure a difference in potential, so they need two connections, one either side of the device under investigation. Voltmeters only work when the current flows in a particular direction, and they must be connected the correct way round. Without a pd (or voltage value) between two points in a circuit, no current will flow.

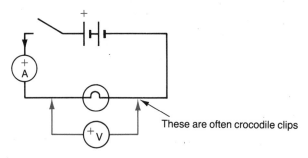

These are often crocodile clips

Fig. 8 Voltmeters are always connected in parallel

Source of pd

When you choose a source of electricity, you fix the pd value. For example, the pd value of a single cell is approximately 1.5 V. This cannot change and remains at this value whatever circuit you put it in. Laboratory power packs have several pd values to choose from, but once chosen, do not change (unless you make the change).

Two 1.5 V cells in a series add together to provide a pd of 3 V. Three cells would become 4.5 V etc.

Sharing pd around a circuit

The pd value between the two terminals of a battery provides the driving force to send a current completely around the circuit. Each component will require some part of the (total) driving force. Each component will have a pd between its two connections. The sum total of all the individual pds will equal the pd of the battery. The circuit diagram in Fig. 9(a) shows an example of this.

(a)

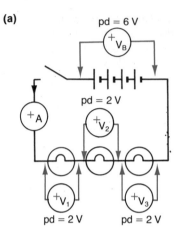

(b)

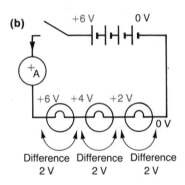

(c)

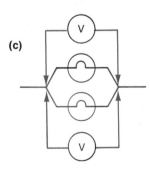

Fig. 9 (a) The sum of all pds equals the pd of the battery. (b) Potential difference decreases around the circuit. (c) Both voltmeters will show the same value, no matter what value resistance the bulbs are.

The example in Fig. 9(a) uses a battery with a pd of 6V. That is, the (+ve) terminal has a value of +6 V and the (−ve) terminal has a value of 0V. The difference between the terminals is 6V. The potential around a circuit decreases in this example in steps of 2 V. Fig. 9(b) shows how the pd value decreases from +6 V to 0 V. The pd values for a parallel circuit need a special mention. Fig. 9(c) shows a pair of voltmeters connected to the ends of the parallel pair. They are both connected to the same places and so they will both show the same value. You could do the same thing by using one voltmeter and taking two separate readings. The pd readings are the same for each member in a parallel section of a circuit.

Potential divider

A potential divider does exactly what its name suggests. It divides potential. It is a resistor with a 'roving' or movable connection somewhere in the middle.

The diagrams in Fig. 10 show how one is used. (They are sometimes called *potentiometers*, or 'pots'.)

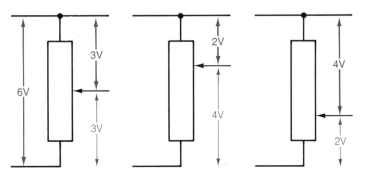

Fig. 10 Potential dividers

Ohm's law

There are three measurements which can be made for any electrical circuit: current in *amps*, potential difference in *volts* and resistance in *ohms*. Ohm's law links the three together. Consider these statements:

If the resistance in a circuit increases, the current decreases. If the pd is increased the current will increase.

Ohm's law states that:

$$\underset{\text{(volts)}}{\underset{V}{\text{Potential difference}}} = \underset{\text{(amps)}}{\underset{I}{\text{Current}}} \times \underset{\text{(ohms)}}{\underset{R}{\text{Resistance}}}$$

$$V = I \times R$$

Alternative arrangements are:

$$R = \frac{V}{I} \quad or \quad I = \frac{V}{R}$$

A very simple demonstration of this can be done with the circuit in Fig. 11. The variable resistor allows you to share out the

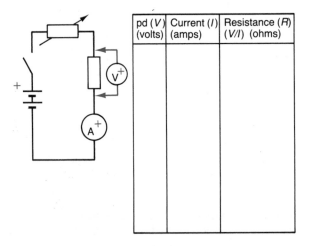

pd (V) (volts)	Current (I) (amps)	Resistance (R) (V/I) (ohms)

 Fig. 11 Finding resistance

pd available between the resistor and itself. By doing this you can alter the current flowing in the circuit. With the slider in different positions, take several pairs of readings of current and pd. Use Ohm's law to calculate the resistance of the resistor. The values should be reasonably constant.

If the resistor becomes very hot during the practical, the resistance value calculated in the third column will increase. Plot the current/pd values on a graph similar to the one in Fig. 12.

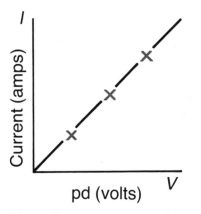

Fig. 12 Plotting current/pd values

Current/pd graphs

It is necessary to be able to interpret current/pd graphs and their connection with resistance. A typical graph obtained from a filament lamp (Fig. 13(a)) is not completely straight, but begins to curve at high voltage values. The resistance is increasing at this point. For many materials resistance increases with temperature (bulbs get hot).

Devices that obey Ohm's law give straight line graphs (Fig. 13(b)). The slope of the graph indicates resistance. A steeply sloping graph would be produced by a low resistance and a more gently sloping graph would be produced by a high resistance.

(a)

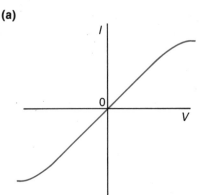

(b)

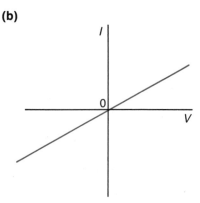

Fig. 13 Current/pd graphs. (a) Filament lamp. (b) Device obeying Ohm's law

Alternating current (ac)

Direct current from a battery may be made to alternate by removing the connection, turning the battery around and then reconnecting it back in the circuit. The current will now be flowing in an alternative (opposite) direction. *Alternating current* such as household mains alternates (undergoes a complete change) 50 times every second. It has a frequency of 50 hertz (symbol, 50 Hz ac or 50 Hz ∼). It is made to do this at the power station.

Comparing ac with dc

Using a dc meter with an ac source will not usually work and may result in damage. There are meters that can be used with both ac and dc, but even these will not tell you much about quickly changing alternating supply. A more practical tool to use is the *cathode ray oscilloscope/tube* (CRO/CRT). CROs are very fast-acting voltmeters. They work quite happily with ac and dc supplies.

Using a CRO

You should be familiar with the operation and use of a CRO. Fig. 14 shows the layout of a typical CRO with basic controls. The controls are used to adjust and control the beam or trace which crosses the screen.

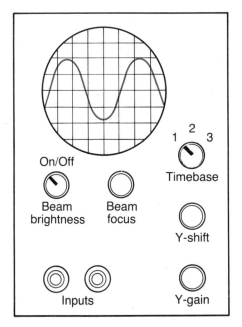

Fig. 14 Typical CRO layout

On – off/beam brightness
On some CROs these functions are separated. The brightness
control increases or decreases the brightness of the trace. Once
adjusted it should need no further attention.

Beam focus
The trace should be focused when the CRO is initially switched on
and should require no further attention.

Timebase
By using the timebase, the speed of the trace may be controlled. If
the passage of the trace across the screen is speeded up it will
take a shorter time to go from one side to the other. With a fast
timebase it is possible for the trace to 'trace out' events which
happen in a very short time. With the timebase set to a lower
speed, the trace will take longer to cross the screen and slower
events can be observed.

Y-shift
The Y-shift's only function is to move the entire trace up or down
the screen.

Y-gain
The 'Y' direction is up/down. The word 'gain' suggests an
increase. This control increases/decreases the size of the trace in
the Y direction only.

Input
The two input terminals are usually coloured black/red. If you
make the connections the wrong way round, the trace will be
upside down.

The grid pattern on the screen is useful for comparing voltage
values. For example, if 2 squares represents 5 V, then 4 squares
will represent 10 V, etc.

CRO traces
The three diagrams in Fig. 15 are typical of CRO traces. Before a
voltage was applied to the input terminals the trace was set along
the centre grid line. Traces (b) and (c) differ only by the height of
the trace. If the CRO controls were not altered then (b) would
represent 6V ac. If the input to (c) was not changed, then the Y-

gain control would have had to have been altered, to give the trace in (c).

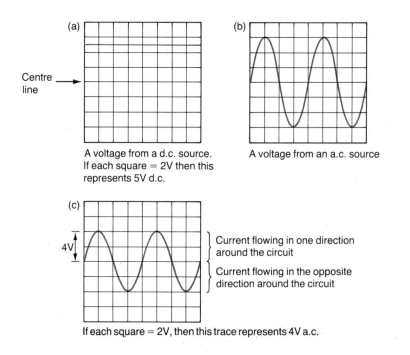

(a) A voltage from a d.c. source. If each square = 2V then this represents 5V d.c.

(b) A voltage from an a.c. source

(c) If each square = 2V, then this trace represents 4V a.c.

Fig. 15 CRO traces

Why ac?

Transmitting electrical energy is most efficiently done using very high voltages and by keeping the current values as small as possible. Unfortunately very high voltages are not practical for everyday use in homes/offices, etc. A way of transforming these very high voltages down to a safer, more convenient value is required. This is completed by using a device known as a **transformer**. Transformers only work with ac. They do not function with dc.

If there was no ac, it would not be practical to transmit electrical power very far; too much would be wasted. Each town or village would need to have its own power station. This is not very practical. Where would the power station fuel come from?

The National Grid system

Electrical energy from the power station, at 22 000 V, is first transformed up to about 275 000 V. This may be further transformed to 400 000 V. It is then fed to the National Grid system. The network of pylons/wires covers the country and lets the power be transmitted efficiently to wherever it's needed (Fig. 16). This may be a long way from the generating station. The electrical energy is received by the local substation which transforms the grid voltage down to 11 000 V. Another series of transformers finally reduce this to 240 V ac for household use.

Household supply

Mains electricity in the UK is supplied at 240 V and 50 Hz ac (50 Hz ~). This is supplied to a main junction box from which various circuits are made available. The total current available to flow into the junction box has a typical maximum value of 60 A. If this maximum current value is exceeded, then the junction box will normally close down (known as *tripping* or *fusing*).

Ring mains

The separate circuits coming from the distribution box are known as ring mains. In a typical house there might be five ring mains:

1 Downstairs sockets.
2 Downstairs lights.
3 Upstairs sockets.
4 Upstairs lights.
5 Electric cooker.

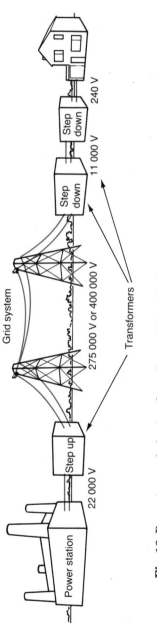

240 V

Step down

11 000 V

Step down

Grid system

275 000 V or 400 000 V

Transformers

Step up

22 000 V

Power station

Fig. 16 Power transmission is only really possible with the use of ac and transformers

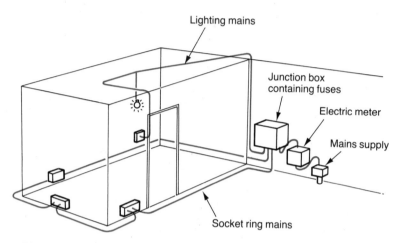

Fig. 17 Household ring mains

Fig. 17 shows part of a typical layout for a downstairs socket ring main/lighting ring main. Ring main circuits are usually buried in the walls or threaded under the floorboards, etc. They are expensive and awkward to replace.

Wire colour coding

There are *two sets of colour codes* for dealing with household electricity. The mains cables that supply electricity within the house, i.e., ring mains, are colour coded:

Red	live
Black	neutral
Bare copper	earth

Appliances, such as toasters, have a different colour code:

Brown	live
Blue	neutral
Green/yellow	earth

Live, neutral and earth

Do not assume that the neutral wire is safe, even though its name may suggest safety. The *live cable* in the household mains is made to 'push and then pull' the electricity around the circuit. The *neutral wire* only completes the circuit each time. In effect, the live cable is the source of electricity all the time. For this reason fuses and switches are always placed on the live side of a circuit.

The *earth wire* is a safety device. Suppose the live cable inside an appliance were to accidentally touch the outer casing. Anything that then touches the casing could receive an electric shock. The earth connection, which is specially connected to the outer casing, provides an easy alternative path for the electricity. The earth connection/wire is an important safety measure, particularly with appliances requiring sockets.

Plugs

It is important you know how to wire a plug correctly and safely. Make sure you practise this. Fig. 18 shows the internal arrangement of a typical three-pin plug.

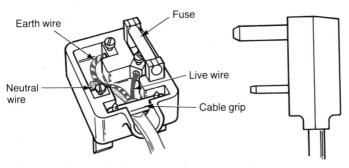

Earth wire

Fuse

Live wire

Neutral wire

Cable grip

Fig. 18 Household electric plug

When wiring a plug, you *must* follow these safety procedures:

1 Do not use frayed wire – trim the end neatly.
2 Do not use excessive lengths of wire.
3 Do not use a fuse of the incorrect value.

4 Do not use wires with split or cracked insulation.
5 Do not attach more than one appliance to a plug.
6 Do not forget to use the cable grip on the outer cable insulation.

Electrical energy and electrical power

As stated in Chapter 2, energy and power are not the same. They are very closely linked!
  Electrical energy is measured in joules (as are all other energy forms). Three factors are involved with the calculation of electrical energy. Increase any one of these and the total amount of energy used will increase: pd (volts); current (amps); time (seconds).

$$\text{Electrical energy } E \ (\text{joules}) = \underset{(\text{amps})}{I} \times \underset{(\text{volts})}{V} \times \underset{(\text{seconds})}{t}$$

$$\text{Electrical power (watts)} = \frac{\text{Energy (joules)}}{\text{Time (seconds)}}$$

$$= \frac{I \times V \times t}{t} = I \times V$$

We can summarize this:

Energy $E = I \times V \times t$ (joules)
Power $P = I \times V$ (watts)

Example Find the power output of a power pack that drives a 12V electric motor by supplying a 3 A current. How much energy will be consumed in 10 s?

To find the power output:
$$P = I \times V = 3 \times 12 = 36 \text{ watts}$$
Power output $= 36 \text{ W}$

To find the energy output in 10 s:
$$E = I \times V \times t = 3 \times 12 \times 10 = 360 \text{ joules}$$
Energy used $= 360 \text{ J}$

A machine that has a power rating of 1 watt is not very powerful. More convenient and meaningful units need to be used:

1 kW	= 1000 W (= 1000 J/s)
1 hour	= 3600 s

More conveniently (for household purposes):

Electric power is measured in kW
Electrical energy is measured in kW hours

Costings by the Electricity Boards

Example A 1-bar electric fire has a power rating of 1 kW. Find the cost of a day's use (8 hours) at 5 p/hour.

Power rating	Hours used	Cost of 1 kW for 1 hour	Total cost
1 kW	1	5p	5p
1 kW	2	5p	10p
1 kW	8	5p	40p

Electricity Boards make a charge for every kW of power used for every hour switched on. 1 kW for 1 hour is 1 unit (1 kW hour = 1 unit). In this example each unit costs 5p.
   Meter readings are usually made quarterly by the Board.

Example	Meter reading at 3 July 1988	= 80347(units)
	Meter reading at 3 Oct 1988	= 81611(units)
	Units used	= 1264
	Cost/unit	= 5.36p
	Amount owed	= 5.36 × 1264 = £67.75

Fuses/loading

Whenever current flows through a material then the material begins to warm up. Apart from electric fires, the heating effect is not usually a problem. Things become dangerous if: wires get hot that are not supposed to; insulations begin to melt. *A higher current will result in greater heating.*
   Wires and cables have current ratings. They are made to allow certain maximum values to flow through them without

overheating. These values should not be exceeded. To stop wires overheating **fuses** are included in electrical circuits.

Fuses are made to melt when the current flow through them reaches a certain maximum value, which is well below the danger point for the wires. Fuses are labelled in amps. There is a wide range to choose from: 1 A, 2 A, 3 A, 5 A, 10 A, 12 A, 13 A and 15 A are common.

The main purpose of a fuse is to protect the supply wiring, not the device itself.

Each ring main connected to a household *distribution/junction box* is protected by a fuse. Socket ring mains are often called 13 A ring mains. If the total supply in a 13 A ring main exceeds 13 A by very much, the main junction box fuse will 'blow' (melt) and the supply will be cut off.

The correct fuse rating for any particular application may be found either directly, by knowing the current value, or by calculating the current value from the power and voltage values.

If you don't know the current value, you can calculate it from the (rearranged) equation:

$$\text{Current} = \frac{\text{Power}}{\text{Voltage}}$$

Example What fuse value is required by a 2 kW mains-operated steam iron?

Power rating = 2 kW = 2000 W
Mains voltage = 240 V

$$\text{Current} \qquad = \frac{\text{Power}}{\text{Voltage}}$$

$$= \frac{2000}{240} = 8.33 \text{ A}$$

The maximum current through the iron will be 8.33 A. From the choice of standard fuses listed previously, there is no exact match, but there are two that are close in value; 7 A and 10 A. If a 7 A fuse were fitted it would blow when the iron was on maximum. The 10 A fuse should be chosen.

For safety, always use the fuse value next above the actual

required; e.g., if the current value is 3.5 A the correct fuse would be 5 A.

There is a simple experiment that illustrates the *heating effect of an electric current* and the use of a fuse. A suitable arrangement is shown in Fig. 19.

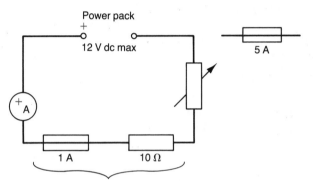

These should rest on a heatproof mat

Fig. 19 This experiment shows the heating effect of electric current

The variable resistor should be set to allow minimum current. By gradually decreasing the variable resistor value, the current will increase. The 10 Ω resistor will begin to warm up. It is unlikely that the 1 A fuse will 'blow' at exactly 1 A, but at a value a little way above. It should 'blow' and stop the resistor overheating. Repeat the experiment with a 5 A fuse. This time the 10 Ω resistor will probably start smoking and may even burn.

Special fuses

Certain appliances require fuses of a specified value regardless of the actual current value through the device. Televisions, for example should always be fitted with a 13 A fuse.

Hazards

Electrical hazards are prevented by common sense:
1 Do not use damaged wires.
2 Do not fit incorrectly rated fuses.
3 Do not overload sockets.
4 Do not take mains-operated equipment into places such as bathrooms.

5 Do not try to 'repair' supply cables.
6 Do not 'poke' anything into a supply socket.
7 Do not, for example, plug a washing machine (requiring 13 A)
into a light socket which may only be capable of supplying 5 A.

Electricity and electrons

This section has been deliberately left till last.

When the early experimenters were first working with simple
electricity experiments such as static, they had no idea what it
was. If two objects created static between them the simplest
explanation was that one object had an excess and the other was
lacking in something, though they had no idea what in.

A material with 'excess' was called 'positive'. A material with a
'deficiency' (shortage) was called 'negative'. Naturally there
would be a flow from positive to negative. Unfortunately in
making this guess of a flow from positive to negative, the early
scientists made a mistake. They had no idea about electrons, and
that they were the cause of it!

Only in 1911 was the electron established as a particle
responsible for the flow of electricity. Electricity is the flow of
electrons. Unfortunately, electrons are negative particles and
flow from (−ve) to (+ve).

Conventional current flows from (+ve) to (−ve)
Electron flow is from (−ve) to (+ve)

This last section can be forgotten unless a GCSE question
specifically asks about electrons/what electricity is. Knowledge of
electrons and electron flow will, however, be useful for further
study in electronics/Physics.

Summary

Static charges are either 'positive' or 'negative'. Materials are
insulators, conductors or semiconductors, and each performs a
different task. An electric current will flow in a complete circuit.
It can be controlled or limited by a resistor. A potential difference
is needed to drive a current through a resistor. Fuses and
switches must always be connected to the live wire. None of the
three wires in a mains cable should be touched. Transformers are
essential for effective power distribution, and only work with
alternating current.

Practice question

Try this question first without reference to the text. Check your answers with the specimen answers at the end of the book. Answers to questions involving calculations should show all the working and should contain a small amount of explanation, including any equations used.

1 (a) Two heating elements (each 15 Ω) are connected in series to a 12 V car battery. What is the total resistance in the circuit and what current flows through them both? Suggest a suitable fuse value.
(b) The same two heaters are now connected in parallel. What would you need to do with the fuse and why? (3/4 sentences).

6 Electronics

Aims of the chapter

By the end of this chapter you should be able to:

1 Recall the varieties/uses of electronic switches.

2 Distinguish between:
   (a) electronic components;
   (b) logic gates.

3 Describe experiments showing the effect of various electronic components.

4 Demonstrate some applications of electronic components/logic gates.

Electronics is switching

There are two types of electronic switches: on/off switches, e.g. light switches; so-called 'dimmer' switches, e.g. volume control. Most electronic devices fall into these two groups.

Switches in the on/off group are used to start or stop something. Switches in the 'dimmer' group will turn something up or down.

Electronics from electricity

Electronics deals with the study of *switching devices* that require electric currents to make them work. Electronic devices are made using *conductors, insulators* and a new material: **semiconductors**. Semiconductors are materials that act like insulators but that can be 'switched on' to act as conductors. The *semiconductor diode* is one such device.

Voltage rails/lines

Rather than keep on using phrases like 'connect a battery', etc., it is far easier to use *voltage rails*. These are a pair of connections rather like railway lines. One of the lines or rails is connected to the positive terminal of a source (of electricity) and the other rail is connected to the negative terminal (of the same source). Fig. 1 shows two examples of a pair of voltage rails. Fig. 1(a) shows a connection between the rails. The flow of electricity here is along the top positive rail through the light emitting diode (LED), on through the resistor, returning along the bottom negative rail. The LED will become switched on. The circuit in Fig. 1(b) shows two connections between another pair of voltage rails. (Note that the one on the left-hand side will be switched on but the one on the right-hand side will not work because the LED is the wrong way round.)

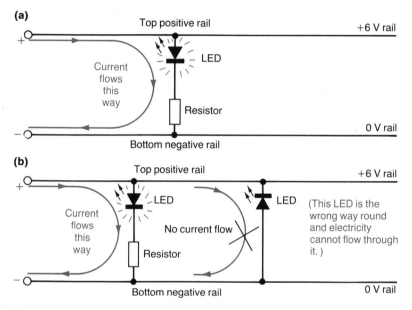

Fig. 1 Positive and negative rails

Semiconductor diodes

These are one-way (switch) devices. Current flows very easily through them in one direction, the direction of the arrow head (known as **forward bias**). In this direction there is a very low resistance. Connecting a diode or light emitting diode the wrong way round (known as **reverse bias**) has a totally different effect. Diodes have a very big resistance in this direction and virtually no current flows through them. Fig. 1(b) shows one LED connected with forward bias and one connected with reverse bias.

If too much current flows through a diode or an LED it will become overloaded and may destroy itself. To stop this happening a *protection resistor* (called a *limiting resistor*) is included. They're not needed when the diode/LED is in reverse bias. LEDs are useful indicator lights.

Changing ac into dc (rectification)

Diodes are one-way devices and can usefully be used to convert ac into dc. The traces in Fig. 2 show the difference between an ac supply and a rectified supply.

(a)   **(b)**   **(c)**

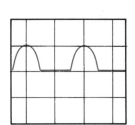
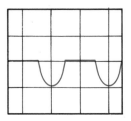

ac supply   Rectified ac using a diode   As (*b*) but with the diode the other way round

Fig. 2 Using a diode to rectify an ac supply

Electronic devices

Light-dependent resistors (LDR)
Light-dependent resistors (LDRs) are 'dimmer' type switches *controlled by light*. A brighter light falling on an LDR allows more current to flow through it. In darkness, an LDR hardly conducts at all. Fig. 3(a) shows an LDR being used as a light meter.

LDRs are very useful for controlling the current flowing into other electronic switches, e.g. transistors.

Thermistors
Thermistors are heat-sensitive resistors. They are 'dimmer' type switches *controlled by heat*. Thermistors conduct electricity more easily when they are hot. They are poor conductors when they are cold. Fig. 3(b) shows a thermistor being used as a thermometer.

Thermistors are very useful for controlling the current flowing into other electronic switches e.g. transistors.

Reed switches
These have nothing to do with reeds. They are switches *activated by magnetism*. They contain a pair of contacts which are

(a)

(b)

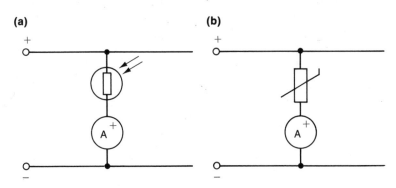

Fig. 3 (a) An electric 'light meter'. (b) An electric 'thermometer'

permanently open – no current will flow. When a magnetic field passes close by, the contacts touch and switch the circuit on. They separate when the magnetism is removed. The magnetism may be supplied by a magnet or an electromagnet.

Relay (switch)
Relays are electromagnetic switches. Fig. 4 shows the arrangement of an electromagnetic relay. A current flowing in the coil causes the soft iron to pivot. This action pushes together the two contacts. These two contacts belong to a separate circuit and so this second circuit becomes switched on when the relay is operated.

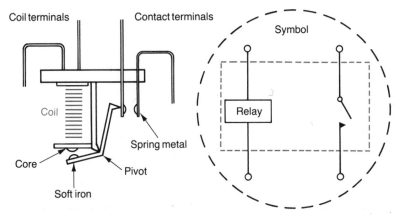

Fig. 4 Relay coil and its symbol

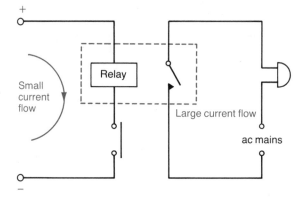

Fig. 5 A very crude burglar alarm

Fig. 5 shows how a small current flowing in the left-hand circuit can control a much larger current in the right-hand circuit. The disadvantage of this arrangement is that the bell will turn itself off when the push-to-make switch is released.

Transistors
These are some of the most useful electronic switches. They have three connections, or legs. They are made in such a way that the *third connection acts as a switch for the other two* (see Fig. 6). If no current flows into the third leg (called the **base**) no current

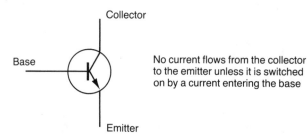

Fig. 6 Diagram of a transistor

flows between the other two (between the **collector** and the **emitter**). The circuit in Fig. 7(a) will not work because no current flows into the base. The LED will not light up.

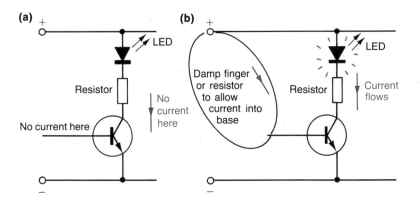

Fig. 7 A transistor circuit (a) open, (b) closed

Just a damp finger (or a resistor) laid between the positive rail and the base connection will switch the transistor on and the LED will light up (Fig. 7(b)). Note: the LED is connected with it's protection or limiting resistor.

Other switches such as the LDR, the thermistor, the reed switch etc. can be used to control the base connection of a transistor. In this way a circuit can be switched on by light, heat, magnetism etc.

Control circuits

Fig. 8 shows some control devices using two electronic switches. For example, in Fig. 8(a) switch No. 1 is the LDR. This activates switch No. 2 – the transistor – which in turn activates the LED.

Each of these circuits switches an LED (and protection resistor) on/off, but it could easily be replaced by a buzzer, a bell, an electric motor, etc.

In every circuit it's the current flowing into the base that controls the transistor. The variable resistor controls how much light, heat, moisture, etc., switches the current on/off.

The next stop on from the circuits in Fig. 8 is to leave them switched on when they have been triggered. This can be done easily by replacing the transistor with a thyristor. Thyristors (sometimes called silicon-controlled-rectifiers – SCRs) are not included in GCSE syllabuses, but are mentioned here for completeness as some of the applications of the circuits so far described only make any real sense if they remain on until they

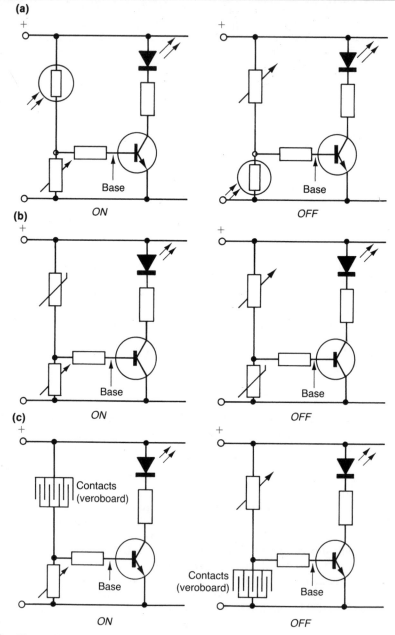

(a)

ON

OFF

(b)

ON

OFF

(c)

Contacts
(veroboard)

ON

Contacts
(veroboard)

OFF

Fig. 8 Circuits controlled by (a) light, (b) heat, (c) moisture

are switched off manually, e.g. light-activated burglar alarm using the circuit in Fig. 8(a).

Truth tables

Fig. 9(a) shows a very simple circuit. When the switch is closed the LED lights up. A *truth table* is a way of describing whether something is 'off' or 'on'.
OFF = 0
 ON = 1

(a)

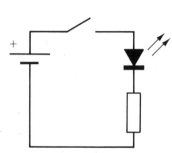

(b)   Switch   Switch
              A   B

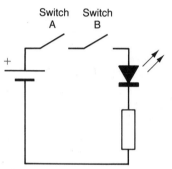

Fig. 9 Circuits for truth tables

The OFF/ON table for the circuit in Fig. 9(a) is:

Switch	LED
Off	Off
On	On

The truth table for the same circuit is:

Switch	LED
0	0
1	1

The truth table for the circuit in Fig. 9(b) would be:

Switch		LED
A	B	
0 (off)	0 (off)	0 (off)
0 (off)	1 (on)	0 (off)
1 (on)	0 (off)	0 (off)
1 (on)	1 (on)	1 (on)

Logic gates

Logic gates are switches. Knowledge of truth tables and some of
the applications for two-input logic gates are all that is required.
Details of their internal workings are not required.

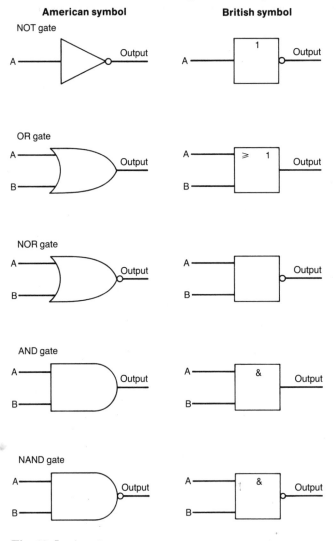

Fig. 10 Logic gates

The five logic gates listed in Fig. 10 and Appendix 1 are shown as both American and British symbols. It would be advisable to cross out those not required and change the symbols in Fig. 11.

AND gate
Both inputs to the AND gate need to be on for this switch to work.

AND truth table   OR truth table

Inputs		Output	Inputs		Output
A	B		A	B	
0	0	0	0	0	0
0	1	0	0	1	1
1	0	0	1	0	1
1	1	1	1	1	1

OR gate
If either of the two inputs are 'on' in the OR gate, the switch will work.

NOT gate (invertor)
This is the easiest switch of all. The output is always the opposite of the single input.

NAND gate
The NOT-AND (NAND) gate switches on unless the 'AND' combination is present at the two inputs.

NOT truth table   NAND truth table

Inputs A	Output	Inputs		Output
0	1	0	0	0
1	0	0	1	1
		1	0	1
		1	1	0

NOR gate
This NOT-OR (NOR) gate only switches on when both inputs are off.

NOR truth table

Inputs		Output
A	B	
0	0	1
0	1	0
1	0	0
1	1	0

Logic gate application

Fig. 11 shows a thermostatically controlled water heater. The thermistor acts as a thermometer and the variable resistor sets the temperature at which input A is 'on'. The AND gate will not work unless input A and input B are both 'on'. Input B will only be 'on' if electricity is conducted between the bared contacts. When the temperature setting is reached *and* the contacts are covered with water, the AND gate switches on. It switches on the relay and the heater in the second circuit warms up the water.

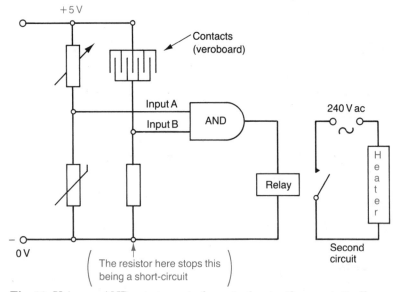

Fig. 11 Using an AND gate to control a water heater thermostatically

Summary

Electronics is about switching. Components, including semiconductor devices, are often arranged between voltage rails. Electronic circuits make useful control devices. Truth tables are used to predict how a set of logic gates (switches) will operate.

Practice question

Try this question first without reference to the text. Check your answers with the specimen answers at the end of the book.

1 (a) The circuit in Fig. 12 is designed to switch on a set of anti-burglar lights when the light level goes down. Place the three components into their correct places.
(b) Why bother with two circuits? Why not just have one? (3/4 sentences).

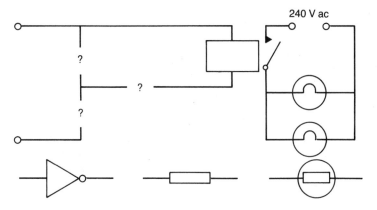

Fig. 12 What goes where?

7 Magnetism

Aims of the chapter

By the end of this chapter you should be able to:

1 Recall:
(a) that there are two magnetic poles and the test for magnetism is repulsion;
(b) that the earth acts as a huge bar magnet and that a magnetic field has direction;
(c) the right-hand-grab-rule can be used to show the direction of an electromagnetic field.

2 Distinguish between magnetically hard and soft materials.

3 Describe experiments to:
(a) plot the magnetic field pattern around a source of magnetism;
(b) identify the polarity of a magnetic source;
(c) illustrate magnetic induction;
(d) show the magnetic effects of an electric current in a straight wire and a solenoid.

4 Demonstrate a practical knowledge of a selection of magnetic and electromagnetic devices.

Simple magnets

At its simplest, **magnetism** is an effect of one thing on another. A **magnet** is a source of magnetism. Around all sources of magnetism is an area where magnetic effects happen. The area in which these effects are noticed is known as a magnetic field. It is sometimes referred to as a **magnetic field pattern**. Some materials make better magnets than others. Materials containing steel, iron, nickel and cobalt make some of the best magnets. Steel is very good at keeping its magnetism and because of this it is known as a *magnetically hard material*. At the other extreme, iron can lose its magnetism very easily. Iron is a *magnetically soft material*.

Magnetic poles

The magnetic field around a simple bar magnet can easily be shown by laying a piece of paper over the top of the magnet and then sprinkling this with iron filings (see Fig. 1).

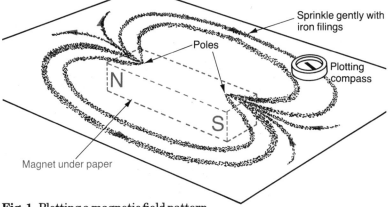

Fig. 1 Plotting a magnetic field pattern

The pattern in Fig. 1 clearly indicates a concentration of magnetism near the two ends of the magnet. Following the pattern back into the magnet leads you to where the magnetism 'appears' to start from. These positions are known as **magnetic poles**. They always appear in pairs. One will be a *north-seeking pole*, the other will be a *south-seeking pole*. For convenience a magnetic field pattern is given a direction, from north to south. This is indicated by arrowheads.

The earth as a magnet

If magnetic poles are free to swing then a magnetic north-seeking pole will always point in a northerly direction and a magnetic south-seeking pole will always point in a southerly direction. A compass is a small free-swinging magnet. If the magnetic north pole swings in the general direction of the north pole then a force must be at work to make this happen. A magnetic north pole is attracted to a magnetic south pole. The earth is thought to act as if it had a huge bar magnet inside it. Surprisingly, it has a magnetic south pole under its geographic north pole (Fig. 2).

Magnetic induction

If a steel paper clip is brought near the pole of a magnet it will be attracted and appear to cling on. A second paper clip can usually

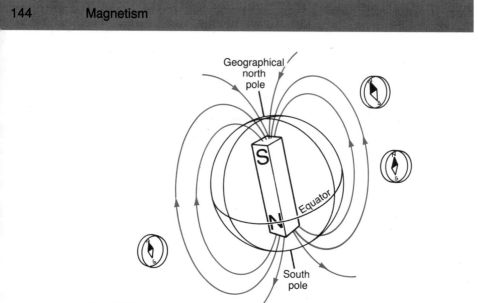

Fig. 2 The earth as a magnet

be suspended from the first, even though the first is not a magnet. This happens because the first clip allows the magnetism (from the magnet) to pass on through it. It is quite possible to continue this and hang several paper clips from the first. Each clip in turn passing on some of the magnetism. The idea of magnetism transferring from one object to another is called **magnetic induction**. For magnetic induction to occur it is not necessary for the materials to be touching. They can be separated, for example, by paper or air. It is by the method of induction that magnets are able to attract non-magnets (see also Electromagnetism, below).

Attraction and repulsion

If two like poles (north-north or south-south) are brought together the effect is that of repulsion. They will want to push themselves apart. If two unlike poles are brought together the effect is that of attraction. There is one other possibility. A magnet will also attract a piece of material by magnetic induction. The only true test for magnetism is *repulsion*.

Electromagnetism

Magnets are not the only source of magnetism. Whenever electricity passes along a wire, a circular **electromagnetic** field is produced around the wire (Fig. 3(a)). If direct current is used then the electromagnetic field will be in one direction only. This is easily checked with a small compass. Hang a current-carrying wire vertically. Hold the compass horizontally and bring it close to the wire. By moving the compass slowly it is possible to 'follow' the north pole. It will lead you in circles around the wire. The *right-hand-grab* rule is used to show the direction of the magnetic field. If the wire is coiled into the shape of a spring it is given the name **solenoid**. The magnetic field pattern around a solenoid is the same as that of a bar magnet, and it behaves in the same way. It has a north-seeking pole and a south-seeking pole. This, too, can be checked with a small compass. Gripping the solenoid with the right hand will let you correctly identify the poles of the solenoid (by the right-hand-grab rule; see Fig. 3(b)).

(a)

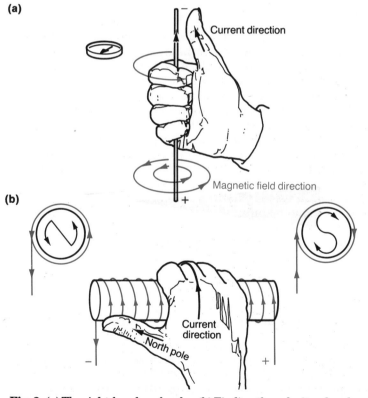

(b)

Fig. 3 (a) The right-hand-grab rule. (b) Finding the polarity of a solenoid

Placing a (magnetically) soft iron core inside the solenoid creates a more practical electromagnet. The magnetism disappears as soon as the current is switched off. (If alternating current is used the magnetic field will alternate too. It will alternate at the same rate as the electricity. If a 50 Hz alternating mains supply is used the magnetic field will undergo a complete change 50 times a second.

Recording heads

The recording head in a tape recorder is an electromagnet. When sound is converted to an electrical signal this switches the recording head 'on/off'. The electromagnetism created by the head causes the tape to become magnetized.

The transformer

Transformers allow electrical voltages to be changed up or down very easily. Without transformers, many of the things that rely on electricity would not be possible. A particular use is in the transmission of electrical power via the National Grid. A transformer consists of a central core around which two coils of wire are wrapped. The coils will usually have different numbers of turns. The core is often made of (laminated) soft iron. One coil is connected to the terminals of an alternating power supply. This creates an alternating magnetic field inside the coil, in the same way as it would in a solenoid. The second coil is affected by the changing magnetic field. In this coil the reverse effect happens. Magnetism is converted back into an alternating electric current. The energy changes are:

K

Electricity → Magnetism → Electricity
(coil 1)   (soft iron core)   (coil 2)

A transformer will not work with direct current.

Movement and magnetism

The force of attraction or repulsion created by a pair of magnetic fields is not limited to ordinary magnets. A pair of electromagnets or one bar magnet and an electromagnet will affect one another in the same way. The controlling of this attraction or repulsion is the basis of machines such as electric motors.

Electromagnetic machines

The following devices all rely on attraction/repulsion by one or more magnetic fields.

The electric bell
When the electric current (see Fig. 4) is switched on, the iron core becomes magnetized. This attracts the iron clanger towards it. The bell sounds. This movement also disconnects the electric supply at the contact point. The current stops and the electromagnetism disappears. The spring steel returns the clanger to its original position. This restores the electrical contact. The process starts all over again.

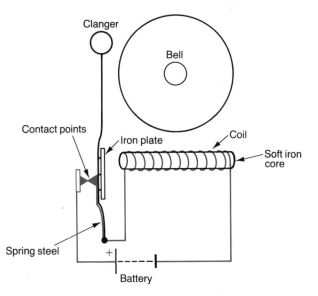

Fig. 4 An arrangement for an electric bell

The moving coil meter
When an electric current passes through the coil (Fig. 5, overleaf) it produces electromagnetism. This reacts with the magnetism from the permanent magnets placed on either side of the coil. The only movement possible is rotation. The coil turns. The return spring controls how much turning can take place. The greater the current flowing through the coil the greater will be the turning

effect. The pointer will indicate a higher current value on the
scale. It is important to connect the terminals of a moving coil
meter correctly, otherwise it will try to run in reverse.
Alternating current should not be used with this device as it may
damage it. With suitable scales, moving coil meters may be made
into voltmeters or ammeters. Other practical uses include fuel
and temperature gauges.

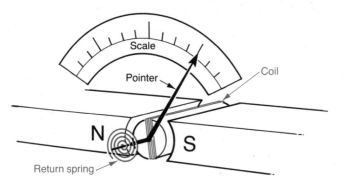

Fig. 5 The round ends to the magnetic poles provide a more uniform
field

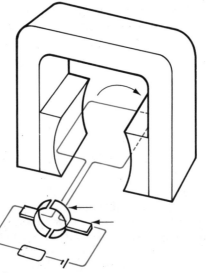

Fig. 6 Simple dc motor

The dc electric motor

This works in basically the same way as the moving coil meter. There is no return spring. The motion is continuous and in one direction only (see Fig. 6). If the terminals or magnetic poles are reversed the direction of motion will also be reversed. The motor can be made to rotate faster by: increasing the strength of the permanent magnets; increasing the number of coils; or increasing the current flowing in the coils.

The dynamo/generator

An electric dynamo/generator is an electric motor running in reverse. Instead of using electricity to create motion, motion is used to create electricity. The generator can be made to produce more electricity by: increasing the strength of the permanent magnets; increasing the number of coils; or increasing the speed of rotation of the coil.

Loudspeakers

A small coil is suspended inside a large circular magnet (see Fig. 7). It usually has the speaker cone attached to it. When an electric current passes through the coil, electromagnetism is produced. This reacts with the magnetism from the circular magnet. The movement produced is rather like that of one tube trying to move inside another, either backwards or forwards. As the cone is attached it moves too, producing sound.

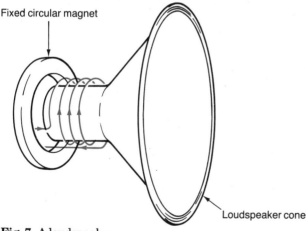

Fixed circular magnet

Loudspeaker cone

Fig. 7 A loudspeaker

Summary

Magnets can affect other materials in their magnetic field by the force of attraction or repulsion. Some materials retain their magnetism permanently, others do not. The earth acts like a huge bar magnet. Electromagnets create the same effects as ordinary magnets; electromagnets can be switched off, but ordinary magnets can't. Electromagnets are used in many common devices.

Practice question

Try this question first without reference to the text. Check your answer with the specimen answers at the back of the book. Limit your reply to about the length indicated in the brackets.

1 An electromagnet is made by winding a coil of wire on a soft iron rod, as in Fig. 8.
(a) When a current flows in the coil what happens to the soft iron rod? (2 sentences)
(b) Why is steel not suitable for use as the core of the electromagnet? (2 sentences)
(c) Explain why the lamp lights when the switch S is closed. (3 sentences)

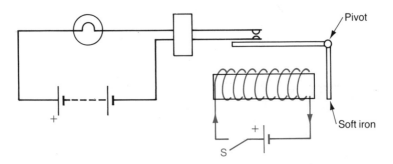

Fig. 8 An electromagnet

Circuit symbols

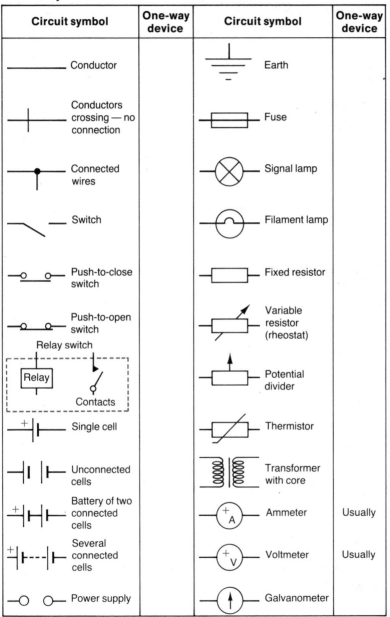

Circuit symbol	One-way device	Circuit symbol	One-way device
Conductor		Earth	
Conductors crossing — no connection		Fuse	
Connected wires		Signal lamp	
Switch		Filament lamp	
Push-to-close switch		Fixed resistor	
Push-to-open switch		Variable resistor (rheostat)	
Relay switch — Relay — Contacts		Potential divider	
Single cell		Thermistor	
Unconnected cells		Transformer with core	
Battery of two connected cells		Ammeter	Usually
Several connected cells		Voltmeter	Usually
Power supply		Galvanometer	

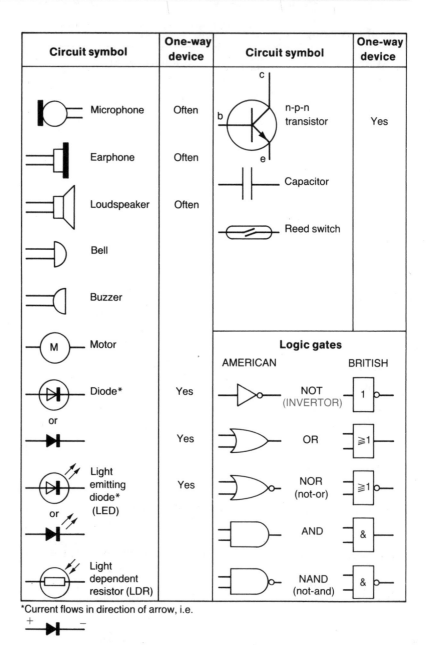

Circuit symbol	One-way device	Circuit symbol	One-way device
Microphone	Often	n-p-n transistor	Yes
Earphone	Often	Capacitor	
Loudspeaker	Often	Reed switch	
Bell			
Buzzer			
Motor		**Logic gates**	
Diode*	Yes	NOT (INVERTOR) — 1	
	Yes	OR — ≥1	
Light emitting diode* (LED)	Yes	NOR (not-or) — ≥1	
		AND — &	
Light dependent resistor (LDR)		NAND (not-and) — &	

AMERICAN　　　　BRITISH

*Current flows in direction of arrow, i.e.

+ ▶◀ −

Mathematical equations

There is no need to remember these questions as they should be provided on the GCSE examination papers. They are included for you to practise using. They are in no particular order of importance. If no variations are shown they are unlikely to be required. Check your own syllabus.

Basic equation	Variations
Moments $Nm = N \times m$	$N = Nm/m$ *or* $m = Nm/N$
Distance $D = S \times T$	$S = D/T$ *or* $T = D/S$
Acceleration $= \dfrac{\text{Change in speed}}{\text{Time taken}}$	
Force $F = m \times a$	$m = F/a$ *or* $a = F/m$
Weight $w = m \times g$	$g = w/m$ *or* $m = w/g$
Pressure $P = F/Area$	$F = P \times A$ *or* $A = F/P$
Kinetic energy $KE = \frac{1}{2} \times m \times v^2$	
Electrical energy $E = I \times V \times t$	
Work $W = F \times D$	$D = W/F$ *or* $F = W/D$
Power $P = \text{Energy}/t$	$E = P \times t$ *or* $t = E/P$
Electrical power $P = I \times V$	$V = P/I$ *or* $I = P/V$
Charles' law $V/T = \text{const}$	
Boyle's law $P \times V = \text{const}$	
Density $D = \text{mass/vol}$	$m = D \times V$ *or* $V = m/D$

Waves $S = \lambda \times f$ $\lambda = S/f$ or $f = S/\lambda$

Ohm's law $V = I \times R$ $I = V/R$ or $R = V/I$

Series resistors $R_{\text{total}} = R_1 + R_2$

Parallel resistors $R_{\text{total}} = \dfrac{R_1 \times R_2}{(R_1 + R_2)}$

Efficiency $= \dfrac{\text{Energy output}}{\text{Energy input}} \times 100\%$

1 Forces

1 The anticlockwise moment (the load) has to equal 25 Nm

Moment $= f \times d$
$25\,\text{Nm} = f \times 0.4$
$f \quad = 62.5\,\text{N}$

2 Moment No. 3 = 1.25 Nm
Any pair of values for force and distance would be satisfactory as long as they multiply together to make 1.25 Nm.
Moment No. 4 = 1.2 Nm. Any pair of values for force and distance would be satisfactory as long as they multiply together to make 1.2 Nm.

3 Distance travelled = Area under the graph.
Area of triangle (for the first 5 seconds) = ½ × 5 × 10
Distance travelled during first 5 seconds = 25 m
   The next 10 seconds of the journey would be represented by a rectangle.
Area = 5 × 10 = 50 m²

2 Energy

1 Air is an insulator. This helps stop heat passing out of the house walls. Plastic foam is a better insulator than air on its own. It also stops heat transfer by convection because the foam cannot move.

2 Heat energy travels by conduction from the water through the metal radiator. This is passed, also by conduction, to the air molecules next to the radiator. The air warms up. Warm air convection currents start up and move around the room taking the heat enegy with them.

3 Matter

1 (a) The same number of molecules will be forced into a smaller space. The pressure will increase.
(b) The molecules will have less space to move in. They must be closer together.

(c) The mass of gas will stay the same even though the volume decreases. The density will increase.

4 Waves

1

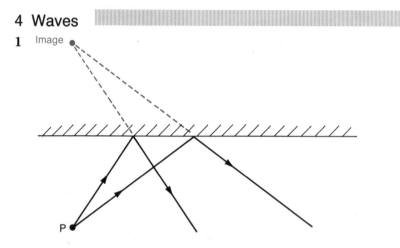

2 (a) Distance = Speed × time
$$= 1400 \times 0.1 = 140 \, m$$
This distance is for an echo (there and back).
Distance to fish = 70 m.
(b) The shoal of fish will not all be at the same depth. The return echo will be a mixture of echos from these various depths. Echos from fish deeper in the water will take longer to return than the echos from fish in shallow water.

5 Electricity

1 (a) Total resistance in a series circuit is:
$$R_{total} = R_1 + R_2$$
$$= 15 + 15 = 30 \, \Omega$$

Ohm's law:
$$I = \frac{V}{R}$$

$$= 12/30$$
$$= 0.4 \, A$$

A sensible fuse to use would be 0.5 A or perhaps 1 A, but no higher.
(b) The total resistance of the pair of resistors in parallel is reduced. The current flowing from the battery would increase and might blow the original fuse. A higher valued fuse would be needed.

6 Electronics

1 (a)

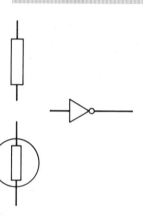

(b) The circuit controlled by the LDR requires only a small voltage. It has only a small current flowing through it and is quite safe. The outside lights are connected to the mains supply (240 V) and have far larger currents flowing through them. This is an example of a small current controlling a much larger one.

7 Magnetism

1 (a) The iron core becomes magnetized. It remains magnetized only when the current is flowing.
(b) Iron loses its magnetism when the current is switched off. Steel is too difficult to magnetize and demagnetize easily.
(c) The iron core becomes an electromagnet and attracts the soft iron armature. The switch contacts will touch. This completes the circuit with the bulb in it.

Last-minute revision should not mean 'the night before'. Revision should be planned. However, if for one reason or another you have only a short while before your exams, there are a number of things you can do to give yourself the best possible chance. The most important of these is not to panic!

Instant revision programme

Go through this book chapter at a time and concentrate on the **key ideas** and **key diagrams**. Examiners require answers to be sensible even if the correct phrase or term has not been used. Test yourself.

If time is really short concentrate on some of the chapters and leave others completely alone. It's better to know some of the work well than know a little bit about everything.

If there's a section you really can't cope with then forget about it. Don't waste time.

Be organized the night before:
- Don't revise too long and too hard;
- Don't stay up too late;
- Don't go to a party or a disco.
- Do arrange for pens, pencils etc;
- Do have an early night;
- Do relax.

Be organized on the day:
- Do have a light breakfast or lunch;
- Do arrive early;
- Do think 'I can do it'.